JURASSIC MARY

To two singular young women,
Amanda
and
Brittany

JURASSIC MARY

MARY ANNING
and the
PRIMEVAL MONSTERS

Patricia Pierce

The
History
Press

Cover illustrations. Front, top: Skull of an *Ichthyosaurus* discovered by Mary
Anning and her brother, *c.* 1811. (© Natural History Museum, London)
Middle: Mary Anning by B.J.M. Donne, 1847. (© Geological Society/
Natural History Museum, London) *Bottom:* Lyme Regis, view from the
west towards the Cobb. (Ian West © 2004)

First published 2006
This paperback edition published 2014

The History Press
The Mill, Brimscombe Port
Stroud, Gloucestershire, GL5 2QG
www.thehistorypress.co.uk

Reprinted 2016

British Library Cataloguing in Publication Data.
A catalogue record for this book is available from the British Library.

ISBN 978 0 7509 5924 7

Typesetting and origination by The History Press
Printed and bound in Great Britain by TJ Books Limited, Padstow

Contents

List of Illustrations

Between pp. 110 and 111.

Chapter opening motif: skeleton of *Ichthyosaurus communis*, found by Mary Anning in 1821 and restored by W.D. Conybeare. (*Penny Magazine*, 7 September 1833)

Box motif: skeleton of *Plesiosaurus dolichodeirus*, found by Mary Anning in 1823 in the shale at Lyme Regis and restored by W.D. Conybeare. (*Penny Magazine*, 7 September 1833)

Title page and chapter tailpieces: Ammonite. (*Author's collection*)

Acknowledgements

Anyone interested in, or writing about, the life of Mary Anning must give thanks and credit to the substantial research carried out by several individuals. One is William Dickson Lang (1878–1966), Keeper of the Department of Geology at the British Museum from 1928 to 1938. His wife came from Charmouth, and he retired there to pursue his studies of local natural history and geology. All the while he became more and more convinced of the value of Mary Anning's pioneering work, and collected as much information as he could at that date. His approach was scientific with references. As president of the Dorset Natural History and Archaeological Society from 1938 to 1940, he published many articles in the Society's *Proceedings*, and these are an invaluable source for all writers and researchers on Mary Anning.

John Fowles (1926–2005), author, novelist, historian and Lyme resident, became Joint Honorary Curator of the Lyme Regis Philpot Museum in 1978, and Honorary Curator from 1979 to 1988. His work on Miss Anning and the history of Lyme importantly focused attention on these subjects.

Today, geologist and historian Hugh S. Torrens, formerly at Keele University, is a leading expert on Mary

Anning. He has undertaken much valuable research, and is currently writing his own full biography of Mary Anning. Particularly useful is his 'Presidential Address' of 1995.

Thanks are also due to Jo Draper, Lyme Regis Philpot Museum; the staff of the Earth Sciences Library, and General Library, Natural History Museum; Dr Jenny Cripps, Curator, Dorset County Museum, Dorchester; and the staff of the British Library. Also to Jaqueline Mitchell, Hilary Walford and Jane Entrican, Sutton Publishing; Marion Dent; Kay Hawkins; Bridget Jones; and my agent, Sara Menguc.

Quotes from *Finders, Keepers* by Stephen Jay Gould, published by Hutchinson, are reprinted by permission of the Random House Group Ltd.

Lyme Regis from the sea.

Introduction

It sounds like the beginning of a fairy tale – the story of a poor cabinet-maker's young daughter who discovered an important and massive fossil at Lyme Regis, Dorset. Mary Anning (1799–1847) unearthed the first complete fossilised skeleton of a 'fish lizard' or *Ichthyosaurus*, when she was about 12 years old. However, Mary's life was no fairy tale, but a struggle against near-impossible odds, although the mystery surrounding some of her life does imbue her story with a certain mythical quality.

In Lyme Regis Mary's future path was set when she was still a girl, and she followed it throughout her life, finding a sequence of some of the earliest palaeontological specimens in the world. For Lyme is situated on an exceptionally fossiliferous coastline, where fossils – the remains or traces of animals and plants preserved by natural processes – were, and still are, to be found in abundance, and often of enormous size. But at that time few people knew what these strange bones and objects were or how they had come to be there.

In her twenties Mary discovered the first complete British *Plesiosaurus giganteus* (1823/4), which became the type specimen (that is, it set the definitions for its category for future reference in identifying further finds). She then found the first British example of the strange winged pterosaur, named *Pteradactylus macronyx* (1828) (renamed *Dimorphodon macronyx*), and then the new species *Plesiosaurus macrocephalus* (1828). That was followed by a strange new genus of fossil fish, *Squaloraja* (1829), another type specimen. There was much more. She was among the first to realise that the small fossils named coprolites found in abundance on the foreshore were actually the fossilised faeces of prehistoric 'monsters'. The huge marine 'lizards' were contemporary with dinosaurs, although some of this story happened before dinosaurs were found, identified as such and so named.

Those professionals who study fossil animals and plants, the palaeontologists, have documented the finds, and it is for such scholars to analyse and explain the specimens in detail. While I am drawn to the diversity and beauty of the geological features of our planet, my interest in Mary Anning is as a woman: an exceptional woman, trapped in the stratified society of the first half of the nineteenth century.

Her achievements were remarkable by any standards, but especially so because she was born and bred in lowly circumstances from which there was little chance of escape. Mary was lower class, female, uneducated, unmarried and a dissenter – one who did not belong to the established Church of England. Lyme Regis was a remote place, its inhabitants socially hampered by the barrier of a strong Dorset accent. This impoverished spinster had to earn her own living, and it was to be in an unusual – and dangerous – way: by finding, excavating and then selling fossils both to casual seaside visitors and to important collectors and museums in Britain and Europe. Any one of her 'handicaps' could have been enough to scupper her chances; however, even though she was not properly recognised – as a socially well-placed man would have been – she did succeed to a large degree.

In spite of every disadvantage, Mary's curiosity, intelligence and industry shone through to such an extent that her story is inexorably welded to the history of fossils found around Lyme Regis, mainly, although not exclusively, of the Jurassic Period, 200 to 145 million years ago.

Researching her discoveries was vital to my understanding of Mary; learning something of her subject and giving rein to my interest helped me to appreciate what fired her enthusiasm and determination. I hope the information gathered here to tell her story will similarly inspire the reader to look further, in acknowledgement of her great accomplishments.

A group of extraordinary and multi-talented men touched Mary Anning's story, and some introduction to them is as essential to the understanding of her circumstances as are the very fossils they collected so obsessively. Pioneer geologist Henry De La Beche (1796–1855) founded the Geological Survey of Great Britain; irrepressible William Buckland (1784–1856), an unforgettable character and a founder of the Royal Geological Society, was the first Professor of Geology at Oxford, and became known as the 'Father of Palaeontology'; William Conybeare (1787–1857) first described many of the Lyme fossils; Roderick Impey Murchison (1792–1871) defined and named the Silurian, Devonian and Permian Periods of geological time, and wrote 350 books, reports and papers; Adam Sedgwick (1785–1873) was the first Professor of Geology at Cambridge, a position he held for fifty-five years, and he introduced the term Devonian (with Murchison).

There were others. Gideon Mantell (1790–1852) wrote the important *Fossils of the South Downs or Illustrations of the Geology of Sussex* in 1822, and discovered four of the five dinosaurs known in his time; Charles Lyell (1797–1875) became the leading geologist of the mid-nineteenth century, and published the hugely influential *Principles of Geology*; eccentric Thomas Hawkins (1810–89), a major collector of fossils, sold his truly unique collection to the British Museum; Richard Owen (1804–92) 'invented' dinosaurs by naming this group of large reptiles *Dinosauria*, and it was through his determined efforts that London's Natural History Museum came into being.

Most of these men were either vicars or doctors. They made important discoveries, were members of the Geological Society of London, travelled widely in Britain and Europe, wrote books and have had books written about them. I have merely hinted at their accomplishments here and in the following text. These were the men, and their wives, in Mary's circle of acquaintances, who sometimes became friends and who – to a degree – treated her as a colleague. They are part of her story.

Mary Anning was generous in sharing her hands-on knowledge gained from everyday experience on the foreshore with the eminent gentlemen scholars who came to visit her. Inevitably, they were not always so generous in giving her the credit she deserved, and she became somewhat bitter as they took freely of her work, discoveries and ideas and presented them as their own, seemingly without a second thought, while she continued to live a hard life all her days.

*   *   *

As a child, my box of treasures already contained a Native American Indian arrowhead I had picked up in a freshly ploughed field, a small chip of '1,000-year-old' Pueblo pottery purchased at a museum with funds from my piggy-bank, and the minute nest of a humming-bird. To add to my collection of oddities, on the foreshore of Lake Ontario I found a stone with what looked like a shell in it. I now know it to be an impression of a common *Pecten* shell. I showed the fossil to my father, who was building a stone fireplace. The 6-year-old girl was thrilled when he promptly put it in pride of place above the keystone, where it has remained ever since.

What draws us to fossils? Perhaps it is the jewel-like quality of, say, an ammonite, or perhaps the intriguing orderliness and stark exposition of the skeletal organisation of animals, huge or tiny. Even a child senses that fossils, with the intricate beauty of nature's symmetry, are gifts from another world. And who can resist a science that casts up evocative words like 'coeval', 'antediluvian', 'primeval' and 'primordial'? Or a science that reveals the previously unknown, spectacularly enormous, terrifying and once-living creatures, some the stuff of nightmares?

Mary Anning can be listed among those extraordinary Englishwomen who have defied the constraints of their time and place – women like writer Freya Stark (1893–1993), a Victorian who travelled adventurously by camel in Arabia, or naturalist Marianne North (1830–90), who fearlessly explored the wilderness areas of every continent in the late nineteenth century, painting hundreds of native flora. Mary was unique, but she was also an example of the indomitable amateur who gets on and makes things happen, a type of person still occasionally encountered. In Mary Anning's case, the 'amateur' soon became the consummate professional.

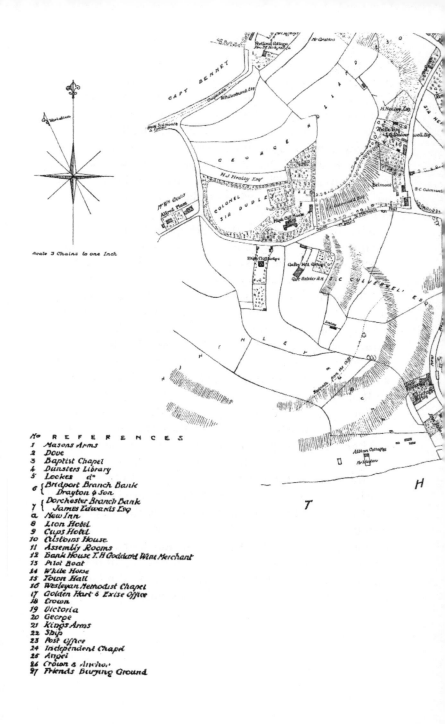

Scale 3 Chains to one Inch

N⁰ R E F E R E N C E S
1   Masons Arms
2   Dove
3   Baptist Chapel
4   Dunsters Library
5   Lockes   d⁰
6 { Bridport Branch Bank
       Drayton & Son
7 { Dorchester Branch Bank
       James Edwards Esq
a   New Inn
8   Lion Hotel
9   Cups Hotel
10   Customs House
11   Assembly Rooms
12   Bank House T. H Goddard Wine Merchant
13   Pilot Boat
14   White Horse
15   Town Hall
16   Wesleyan Methodist Chapel
17   Golden Hart & Exise Office
18   Crown
19   Victoria
20   George
21   Kings Arms
22   Ship
23   Post Office
24   Independent Chapel
25   Angel
26   Crown & Anchor
27   Friends Burying Ground

PLAN
LYME-REGIS
and Environs
1841

1

'Verteberries' and 'Golden Serpents'

There is no picking up a pebble by the brook-side without finding all nature in connexion with it.

(G. Mantell[1])

With the sea breaking behind her and the forbidding cliffs looming in front, the young girl knew she had found what she was looking for. Twelve-year-old Mary Anning, with large intelligent eyes, pale skin glowing in the fresh sea breeze and dark hair tangled, had spotted hints of long bones. A year earlier, in 1811,[2] her brother Joseph had been the first to notice the head of a 'crocodile' in the exposed fallen rocks on the foreshore between Lyme Regis and Charmouth. It would eventually be identified as an *Ichthyosaurus* (literally 'fish lizard'). The huge head had a long snout, and its large saucer-shaped eye socket seemed to stare out at them unnervingly.

Now, one year later, Mary had found the rest of the 'fish lizard' in the cliff high above where her brother had found the head. The Annings hired men to dig out the complete skeleton, 17 feet long, from the place it had rested in for 175, perhaps 200, million years. It was in 'a very perfect state'.[3]

This was the first of these streamlined marine reptiles to receive the full blast of publicity in London. The Lord of the Manor at Lyme Regis, Henry Hoste Henley, bought it from the Annings, and he then passed it on to a museum in London's Piccadilly. But it was not the first ichthyosaur to be discovered. The Welsh naturalist Llhyd had found a few fragments one hundred years earlier, which he described in his book *Lithosphylacii Britannia* (1699), and there had been others. But such a hugely important discovery by this young girl and her brother would prove to be the stuff of legend. This extraordinary find and others to follow would help to challenge, then overturn, many long-held 'truths' regarding the evolution of life on earth.

*   *   *

Mary Anning's birth into a turbulent world was preceded by events of the most dramatic and far-reaching kind: the American Revolution of 1776, the French Revolution of 1789, and the burgeoning Industrial Revolution of 1776–1815, launched in Britain, the first industrialised country. The effects of such world-jarring events even percolated through to life on England's south coast in remote Dorset, which was isolated by muddy, rutted roads, and further set apart by the local dialect.

The Anning family lived in comely Lyme Regis, a small seaside resort in west Dorset, noted for its fossils. Lyme and the River Lym lie cosily in a little combe, or 'bottom', between two steep hills with exceptionally fossiliferous cliffs stretching almost as far as the eye can see in either direction. In a dip to the east snuggles the equally fossil-rich village of Charmouth. Within the crumbling cliffs are strata of Blue Lias, sandwich-like layers of limestone and shale, imprisoning gigantic unknown beasts from other, earlier, worlds. The fossils (mostly skeletal remains, along with wood and trace impressions of prehistoric organisms preserved by natural processes), found almost everywhere around this area of the coast, were millions of years old, dating from periods of 'deep time' that are hard to grasp even today. Generally the fossils were unknown at the time of Mary's discoveries, but some scholars were beginning to understand what they were.

Mary Anning lived out her life against a backdrop of long-recorded history. She was later described as 'a being of imagination'.[4] Did the young Mary gaze out to sea as the mist rolled in, and fancy she had been there at crucial points in Lyme's history? Near the town, *Homo sapiens* had left his marks from the distant past, from the Iron Age hillfort on top of Pilsden Pen, at 909 feet the highest hill in Dorset, and the earthworks at Lambert Castle, to Lyme's very own Cobb, much, much more recent but still hundreds of years old. The Cobb was the central feature of Lyme Regis, and the reason why the little port had survived, even thrived, for centuries.

When searching for fossils, as she walked along under menacing Black Ven cliff to Charmouth, did Mary imagine that she could see the thirty-five fearsome dragon-prowed Viking warships approaching, in AD 831, on one of their frequent raids on that place, 'where they made cruel ravage and slaughter'?[5] Standing on dominant Black Ven cliff, where the warning fire beacon had been

lit on 31 July 1588, could she clearly 'see' the two Lyme
ships heading out into Lyme Bay to join the English fleet
commanded by Drake, and view its first skirmish with
the Spanish Armada?

History flowed all around her. Did she ever imagine
that she had been there in April 1644 during the Civil
War, when Prince Maurice and 6,000 over-confident men
attacked the 'little vile fishing village' intending to make
it 'breakfast work'? For the next eight weeks they
unsuccessfully laid siege to the town from the land side.
Flights of flaming arrows dipped in tar and hot cannon
set ablaze the thatched houses at the west end of the
town, as the intrepid women of fiercely Puritan Lyme
dressed themselves as men to help confuse and repel
the attackers.

When young Mary looked out to sea she might very
well have seen French ships hovering – and she would
not have been dreaming. For centuries there had been
sightings of French vessels, raids and attacks along this
coast. And the Francophobia was firmly grounded on
reality. In addition to the menace of French privateers,
the country was at war with France. When Mary was
born in 1799, Britain had already been at war for six
years and would continue to be so for the first sixteen
years of her life. To a town on the south coast of England,
the wars with the French were not remote. By December
1802 anti-French feeling was high, and between 1803
and 1805 an invasion scare reached panic proportions.
The Napoleonic Wars resumed in 1805 and only came to
an end with the Battle of Waterloo in 1815.

To the west of the Cobb, and taking in the border
between Dorset and Devon, the wild, forested Undercliff,
formed by countless landslides, stretches for 8 miles

between Lyme and Axmouth. With every step Mary could not help but walk over 'serpent-stones' (ammonites), some several feet wide, so many that they formed a pavement, and fossils beyond number on Monmouth Beach. This was the place where James Scott, Duke of Monmouth (1649–85), a natural son of Charles II, chose to land in 1685, unfurl the royal standard and proclaim himself king. When his rebellion failed, the local people paid in blood, and 'Hanging' Judge Jeffreys ordered that twelve men be hanged, drawn and quartered at this spot on the foreshore.

Did Mary ever sense that her story would become an important part of the history of Lyme Regis, or even the most famous?

* * *

Mary's parents, Richard Anning (*c.* 1766–1810) and his wife Mary ('Molly') (*c.* 1764–1842), married at Blandford parish church in August 1793. Richard probably came from Colyton,[6] a market town on the River Coly in East Devon (inland from Beer Head), only 9 miles from Lyme Regis, at a time when the fortunes of the latter were known to be improving. Lyme was becoming a summer seaside resort, attracting well-to-do middle-class visitors.

Mary's mother, Mary ('Molly') Moore, after whom she was named, came from Blandford Forum, a handsome Georgian town not far from Lyme. It was also the name given to an older sister, born in about 1794, who had perished in a Christmas-time house fire in 1798 recorded in the *Bath Chronicle*: 'A child, four years of age, of Mr R. Anning, a cabinet-maker of Lyme, was left by the mother about five minutes . . . in a room where there were some shavings by the fire . . . The girl's clothes caught fire and she was so dreadfully burnt as to cause her death.' Of Molly and Richard's four female infants, three died: Martha, the first Mary and Elizabeth. Of five male babies, four rapidly departed this life: the first Henry, the second Henry, Percival and Richard.

Joseph and his younger sister Mary were the only two children to survive to adulthood out of a family of at least nine children, perhaps ten. In those days it was not unusual for infants to die. The Annings' immediate neighbours in unfashionable, unhealthy Cockmoile Square included John Bennett (1762–1852) and his wife Maria (c. 1763–1831), who lost three infants in four years, also at the end of the eighteenth century.[7] Infant mortality was still high decades later in the first available census (1851): 123 babies died per 1,000 in West Dorset, but this was better than the national average of 150.

Happily, the year 1799 was a healthy one for the newly born in Cockmoile Square where the Annings lived, because not only did (the second) Mary survive, but so too did Ann, born the same year to the Bennetts next door. The Anning house had a double bow window in front, and comprised three floors and a cellar with windows. In the confined area of the Square, close neighbours would have been very close indeed. The Bennetts would remain in the Square for fifty years.

Oddly shaped Cockmoile Square, with its bizarre arrangement of dwellings, was built out almost over the River Lym, with the backs of the houses sensibly facing the sea, but still dangerously close to the foreshore. It was a place for artisans and their families. By trade a cordwainer or shoemaker, John Bennett was at heart a musician, and also an entrepreneur. In the 1820s he opened his own private baths, constructed between his sizeable leasehold property in the Square and the sea, despite the proximity of Jefferd's Baths,[8] open since 1804. That it was a viable proposition illustrates the increasing popularity of Lyme as a seaside resort for the discerning classes.

Richard Anning was a cabinet-maker or carpenter, but he also collected fossils to sell. Clearly a man of independent mind, and with limited time to hunt for fossils, he made use of Sundays and religious holidays including Good Friday and saints' days for his rambles. He was a dissenter, a Congregationalist, as were the Bennetts. The dissenters, too, would have disapproved of his working on Sundays.

Occasionally Richard ferreted out fossils even on weekdays, which angered his wife, who 'was wont to ridicule his pursuit of such things',[9] according to George Roberts, Lyme's schoolmaster and first historian. Money was always short in the endless struggle to feed, clothe and keep the family warm when sharp and penetrating ocean breezes hit the little town tucked so picturesquely in its combe.

Remnants of stories that have come down to us indicate that Richard Anning had a strong, distinctive character. He was one of the ringleaders of a destructive mob during a bread riot in Lyme in March 1800.[10] No one was prosecuted because no one would testify. It was a time, from 1799 to 1801, of bad harvests, when costs spiralled, the price of wheat went up and there were acute local food shortages, especially of the staple bread, since there was no European corn on the market, owing to the war.

DISSENTERS

In Lyme and in all Dorset there had long been a tradition of religious nonconformity. A succession of sovereigns ordered funds to be taken from the Customs at Lyme to maintain the Cobb, a grant withdrawn by Catholic Mary because the 'inhabitants were then reputed as heretics for their religion'.*
Protestant Elizabeth I, who referred to the Cobb as the 'great and sumptuous structure, built with mighty stones and rocks', reinstated the grant.†

In Charles I's reign, the Cavaliers defended the divine right of kings and the State Church of England, while the Puritan Roundheads preferred to worship 'by the word', often with dynamic pastors. On Cromwell's victory, the State Church was abolished, *The Book of Common Prayer* banned and thousands of clergy lost their living. After the Restoration, the situation was reversed and revenge sought. Nearly all the Lyme townsfolk were dissenters, belonging to sects such as the Baptists, Methodists, Quakers and Congregationalists. In nearby Uplyme there were also Anabaptists and Plymouth Brethren.

In Lyme Regis, the Congregationalists met at the Independents' Chapel on Coombe Street. The 'Church Book', dating from 1653, is preserved in a museum in the original building.

Not conforming to the State religion presented an oppressive number of legal liabilities. Dissenters could marry legally only in the State church, which was also the only church from which a valid baptisimal certificate could be obtained. All burial grounds were, initially, owned by the State church; it was also the church to which dissenters were required to pay tithes unless very poor. They were not allowed to attend university and were excluded from many professions. Nor were they permitted to vote until the Municipal Corporations Act of 1835.

* George Roberts, *Roberts's History of Lyme Regis and Charmouth*, London, Samuel Bagster and William Pickering, 1834, pp. 61–2.
† *Ibid.*, p. 226.

The poverty was so widespread in Dorset that people made money any way they could. Even a family's hair was an asset that could be harvested and sold to make wigs and hairpieces; a family's crop of tresses might be farmed out at a price. The barber came along and cut off all the hair, then rubbed the numbskull with oil, and returned when the hair had achieved a marketable length once more.[11] The Annings were fortunate in that fossils were close to hand to provide them with some extra income. From a table in front of his house/shop in the tiny Square, Richard displayed and sold anything interesting or unusual that he or his children had found.

No one knew what the strange, alluring fossils were, so names were invented to explain these 'curiosities', and legends grew around them. Among the variety of colloquially named wares on offer were:

- ❖ 'verteberries' or 'crocodile teeth', believed to come from 'crocodiles' (these were individual ichthyosaur vertebrae)
- ❖ 'petrified serpents', 'snakestones' or 'serpent-stones', thought to have once been serpents (ammonites, an extinct group of molluscs that looked something like the modern squid but with a hard coiled shell). These were the most common fossils and were also known as 'Indies fingers', 'Cornu Ammonis' or 'cornumoniuses' or 'cornemonius', 'arbtusus' and 'birchii', while 'golden serpents' were pyritised ammonites. 'Paper Nautilus' was *Tragophylloceras loscombi*, a smooth ammonite
- ❖ 'ladies' fingers', 'devil's fingers', 'St Peter's fingers', 'arrowheads' or 'thunderstones', believed by some

to be 'thunderbolts' from God (belemnites, an extinct group of molluscs related to ammonites and similar to the modern squid; they had a hard, internal, bullet-shaped skeleton)

❖ 'Angel's wings' or 'Cupid's wings' (masses of marcasite or crystalline iron sulphide)

❖ 'Turbot', 'John Dory', 'John Dory's bones' (*Dapedium politum*, a flat, plate-like primitive fish)

❖ 'Scuttle' (a primitive cuttlefish)

❖ 'Devil's Toenails' (sections through bivalve shells, related to the oyster *Gryphaea*, an extinct oyster-like bivalve mollusc. It had a thick shell, which in fossil form superficially resembles a chunky white toenail)

❖ Fossil shells.[12]

*   *   *

In Mary's day the main entrance to tiny Cockmoile Square was on the seaward side from Bridge Street, hard by Buddle Bridge over the Lym (sometimes called the Buddle), as the river – also the town sewer – nears the ocean at the back of the Square. It sits at the lowest part of hilly Lyme.

The Square itself was a small irregular space – roughly the open space in front of the museum today – with the old town in front and the tempestuous sea very close behind (the ancient sea wall can still be seen). 'Cockmoile' is an early West Country word for lock-up, and indeed the jail stood at the end of the Guildhall on the corner of the entrance to the Square from Bridge Street. In the Square stood the stocks, last used in 1837. According to Lyme's schoolmaster George Roberts,

'Cockmoile' referred to a cock crowing and to labour. Another version has been suggested: 'Cockenwhile', which might be a corruption of 'Coquinaille', meaning a pack of thieves. Everyone seemed to have his or her own interpretation of the name.

To the west of Cockmoile Square was another larger square with the Assembly Rooms on the site of warehouses and the old Cobb Gate in a cramped position on the seafront, and opposite were the Custom House[13] and the prestigious old Three Cups Hotel. Broad Street, the town's steep main street, rises sharply up the hill, and from the top it 'rushes [back] down to the sea', in Jane Austen's words, to meet the square. Opposite Cockmoile Square is narrow Coombe Street, the main street in medieval times, where the plain Independents' Chapel is located,[14] which the Annings attended. There were seven mills on the small River Lym, but Lyme was mainly a town of cottages on hillsides with sloping gardens back and front.

Before the turnpike road was constructed in 1758, thus opening up the village and linking it to Dorchester and Exeter, Lyme was a cul-de-sac. No wheeled vehicles could enter, because of the deep, narrow and muddy lanes and the steep, plunging hills. Goods landed at the port, woollen cloth made by the mills on the river, even fish carried in panniers, all were transported by packhorse teams that gathered by the red-brick George Inn. This medieval inn, where Monmouth himself had stayed so long ago, offered extensive stabling for 'troops of packhorses'.[15] When the inn's gates were shut for the night, it was said that the buildings and activity within resembled a small town. From the George, the packhorses headed up Coombe Street and out to the rest of the world.

Turning right (east) from Cockmoile Square, the road soon makes an abrupt left turn with the Butter Market on the outer corner. Up Church Street stands the Church of St Michael the Archangel, whose six bells can clearly be heard in the Anning house down in the Square. On the way into church the men of a former generation paused to untie the strings of their knee-breeches so they could kneel, and working-class women wearing the whittle – a blanket-shawl dyed crimson – made a 'river of red' flowing out of church or chapel at the end of a service.[16]

From the bend one can turn right into Long Entry towards the seafront. Long Entry leads to a route that ran across the forbidding cliff called Black Ven to Charmouth, an area Mary knew well.[17] The path also led to Gun Cliff, close to the Anning home, where four guns had been sited but were seldom used.

Doubtless Mary Anning would still recognise Lyme Regis today because of the Cobb and 'the remarkable situation of the town',[18] although there have been many changes.

*   *   *

The religious life of Lyme was organised around the church and two dissenting chapels. The three Sunday Schools were 'happily calculated to instruct the ignorant, and improve the lower class of people'. From about 8 years of age Mary attended the Dissenters' Sunday School, where in addition to religious studies she learned the three Rs.[19] Her signed copy of the *Theological Magazine and Review*, Volume 1 for 1801, given to her by her brother, still exists. Interestingly it contains an essay stating that God created the universe in six days, and another essay encouraging dissenters to study geology. On the educational front there were three boarding-schools, two for young ladies and one for young gentlemen, the uncompromising emphasis being on 'ladies' and 'gentlemen'.

Fossiling had long been a tradition in the town. One of two collectors, according to Roberts, was William Lock (c. 1739–1814), nicknamed 'Captain Cury', 'cury' being a local name for 'curiosities'. Fossils had first been exploited at nearby Charmouth, on the main road between Dorchester and Exeter. The bored passengers in the horse-drawn coaches provided a steady stream of new customers for Lock's ammonites and other 'curiosities'.

At Lyme a fossil collector named Mr South (who may have worked for the Duchess of Portland) taught Richard Anning.[20] Another early fossil collector was Mr Cruikshanks (or Crookshanks), whom Richard knew and sometimes accompanied fossiling. His unusual method of investigation and excavation involved prodding the rocks with a long pole 'like a garden hoe' to scratch for hidden treasure. His fate is perhaps indicative of just how precarious a living was to be made from collecting and selling fossils. In 1802, when the coal-merchant business he had left behind in London's Piccadilly stopped sending him an annuity, he found the income from finding and selling fossils was not enough to sustain him, so he killed himself by jumping off Gun Cliff into the sea, only a few yards from Cockmoile Square.

As a child Mary and her older brother Joseph (1795–1849) often accompanied their father on his walks to rootle out fossils, so learning the best places to find these unknown creatures from earlier mysterious worlds came to them as naturally as walking, their young minds absorbing information like sponges. Some may have accused Richard Anning of exposing his children to danger, as no one could deny that fossiling was hazardous work even for an adult.

Richard made his daughter a fossil extractor, a type of hammer[21] with which she was able to release smaller finds from their prison of rock. Later she generously gave her father credit for her mastery of fossils. This was no doubt true of her initial basic knowledge, but her deep understanding of these primeval creatures, largely self-taught, was exceptional, becoming almost instinctive.

Richard Anning's fossiling was noted in a letter written in July 1810 by an early fossil-collector to visit Lyme. The gentleman was James Johnson (1764–1844), writing to another collector, William Cunnington (1754–1810):

> There is a person at Lyme who collects for sale by the name of Anning, a cabinet maker and I believe as men are, may be depended upon, I would advise you calling upon him at Lyme . . . as early as you can spare, you should walk to Charmouth [where James Johnson had once lived] and ask a confounded rogue of the name of Lock to call upon you . . . upon first sight give him a Grog or a Pint, this will buy him to your interest and all crocodiles he may meet with will most assuredly be offered you first.[22]

To reveal the jewel-like beauty of ammonites, it was Richard who reputedly first came up with the idea of slicing them in half to expose the crystalline calcite infilling the chambers.

Those inhabiting the fledgling world of geology formed an intimate group, and Mary was in the right place to get to know most of them. She became acquainted with several eminent pioneer geologists very early on. One future colleague was the boy Henry De La Beche (1796–1855), later to become the greatest geologist associated with the area. Another was Jean André De Luc, almost a Renaissance man so varied were his interests and abilities. They were close at hand from her earliest years, and, in the case of De La Beche, the Annings' fossiling probably drew him into the field in which he would make such a great contribution.

Jean André De Luc (1727–1817) was a highly esteemed Swiss natural philosopher, politician and merchant who had been prominent in turbulent Genevan politics, supporting Rousseau's Social Contract. As a result of the political upheaval and a trade embargo, his business failed.

De Luc arrived in England in 1773, secured a position at the Court of George III, and decided to pursue his love of geological investigation, one of many interests, to which he contributed much valuable research. Geology combined his love of the natural beauty in mountains with his interest in the history of the earth and how it related to man. He published a number of treatises and articles, especially on geology, and was highly regarded throughout his life in both Switzerland and England. In fact De Luc may have been the first to use the term 'geology' in the modern sense.

As part of his research into the character of rock strata, Jean André De Luc began exploring the geology of the Dorset coast in 1805, when De La Beche was a Lyme schoolboy of 11 and Mary was 6. De Luc left an account of a walk from Charmouth to Lyme in which he says: 'It would be difficult to find a country which afforded more agreeable walks' – a compliment indeed from a native of Switzerland. He continued: 'As it was low water, there was a wide strand uncovered below the cliffs, which extend as far as Lyme, a distance of two miles; and along this strand I proceeded.'[23] He noted the changes in strata, the carbonate content of the clays and the inclusions within them, the calcite-filled ammonites and the thin layers of fibrous calcite, known as 'Beef'.

Richard Anning had built up a group of contacts, and was about to add another one. On arriving in Lyme that September De Luc had encountered an unnamed man of whom historian W.D. Lang stated 'there can be no harm in supposing to have been Mary Anning's father'.[24] Lang's description does fit with what is known of Richard Anning. He writes of

one of the inhabitants, who knows this coast very well, because he visits it from time to time, in search of fossils which he sells to the strangers who resort hither; *Lyme* being a place much frequented for sea-bathing. At his house I saw some find [*sic*] *cornu Ammonis*, [ammonites] sawn through the middle and various other marine fossils, proceeding from the above cliffs, and other places.

Together they went to view the 'hill which was crumbling down' by the church, and dangerous to both town and church. They discussed the very narrow space between the church and the disintegrating hill on the seafront, and the man assumed to be Richard Anning said that he could remember the area being much wider and containing gardens. In previous times it had been the promenade, but 'all the ground lost here had detached itself by broad sections, which had successively slidden down the slope to the bottom'. He himself had been there during one of these slips on Church cliffs, and narrowly escaped being carried away with the earth to be crushed to death below.[25]

Unfortunately, Richard's luck did not hold. Late one night in the winter of 1809/10, while walking to Charmouth, he fell off the boggy slopes of the ominous-sounding Black Ven cliff, 'black' because it looked dark and forbidding, 'ven' meaning 'fen' (boggy) in the Dorset dialect, 'having diverged from the path in a field on a summit of the hill'.[26] Perhaps he was unwell, tired or over-confident, because with his experience he should have known better. As one of Europe's largest coastal landslides, the area today is still very unstable, and a dangerous maze if one leaves the paths.

Weakened by the fall, Richard Anning died of consumption in the autumn of 1810, aged only 44. On 15 October 1810 he was buried at St Michael's. Consumption or contagious tuberculosis was associated with deprivation, and accounted for one-quarter of all deaths in Europe at the time. He left his heavily pregnant wife and two children almost destitute, with debts of £120 (almost £3,000 today). A baby, Richard, named after his father, would be born and christened late in 1810, and follow his father to the grave in 1811.

After the death of the head of the household, an already hard life became much more difficult for the family. Their main source of income now was charity. Richard's widow, Molly, applied to the parish for support, and received relief from the Overseers of the Parish Poor – a trifling amount for food and clothing, about three shillings a week, which might have been paid in bread and other staple food. It was not much, but it was something, and they continued to receive these small amounts – which must have caused them great humiliation – until around 1816.[27]

At the time of her father's death Mary was almost 11 and Joseph about 15. It seems Mary did not attend school after this, but her education would continue for a lifetime on the shale of Lyme – a practical, hands-on kind of learning and research – with her intelligent mind absorbing information that cannot be learned from books alone. It would prepare her to become 'the most eminent female fossilist'.[28]

The first major impact that fossils had on Mary's future occurred late in 1810. One day the slim girl was carrying home an ammonite, a coiled fossil that she had found on the beach. She encountered a lady who 'seeing the fossil in her hand, offered her half a crown' for this beauty. Mary accepted. From that moment, full of pride and with money in her hand to give to her mother, she determined to go down 'upon beach' again.[29] This encounter pointed the way to her destiny.

Perhaps this lady had been trying to help the stricken family, as did Mrs Stock, wife of a local landowner, who kept an eye on the Annings. She made sure that they had food, paid Mary to run errands, and gave Mary her first book on geology. A visitor, Miss Anna Maria Pinney, noted in her diary that in Mrs Stock's eyes Mary was a 'spirited young person of independent character who did not much care for undue politeness or pretense'.[30] And widow Molly, desperately poor and with two children to feed, now accepted that selling fossils was a way to survive. The children were eager to help, a task in which they were unusually experienced and knowledgeable.

Just after their father's death, Joseph and Mary found the 4-foot-long head of the famous ichthyosaur in late 1810 or early 1811. Winter was the best season for making new discoveries, as the storms rearranged the surface of the cliffs and the beach, but it was also the most dangerous time. Initially the fossil was believed to be a huge crocodile, the largest reptile then known (marine crocodiles had already been found), with which it did have some similarities. The creature was first described by Sir Everard Home in 1814, then named by Charles Konig, the Keeper of Natural History, British Museum, in 1817, and affirmed by De La Beche from 1818 onwards. It was the most complete specimen of this streamlined marine reptile found to date. Mary's brother, it seems, saw the head first with its huge, long skull and saucer-like eye sockets. The notable ring of bony plates in the eye socket held the eyeball in position and assisted in focusing. Ichthyosaurs, dolphin-like reptiles, had lived at the same time as the dinosaurs, although no one realised this then.

A year later, when the weather allowed, and after fresh surfaces of the cliff-face of Black Ven had been exposed and the foreshore once again rearranged by the corrosive action of the sea, Mary returned to search for, find and excavate the rest of the body of what would become known as *Ichthyosaurus platyodon* (now named *Temnodontosaurus platyodon*). She was undoubtedly helped by 16-year-old Joseph and others. She found the 17-foot marine reptile 30 feet up the cliff, and paid workmen to assist her in removing it.

At this early point in Mary's life, the pioneer geologist and scholar William Buckland heard of her find and immediately sought her out. Thus this great character now entered her life and her story. Buckland had been born in nearby Axminster; his mother was from Lyme and his father was a vicar who collected fossils. Like Mary, his future path had been indicated when a mere child. As a boy he could observe the bones of fossil elephants among tree roots near his home.[31] W.D. Conybeare waxed lyrical in describing what Buckland was like as a boy to Buckland's own son:

> the young Buckland could not take a stroll in the neighbouring fields without stumbling, at almost every step on lias quarries, and finding on ascending every hill, that its summit consisted of an entirely dissimilar formation – chertsand. If he extended his rambles to the shore at Lyme Regis or Charmouth . . . he must have found himself able to walk for miles over the slabs which the lias protruded into the sea, without placing a foot beyond the numerous circles of the larger varieties of his serpent-stones, and found the belemnites aggregated in thousands in particular portions of the cliff.[32]

Could there have been any doubt in which direction his adult interests would lie?

Buckland was an Oxford scholar and from 1809 a Fellow of Corpus Christi College. He would become the wonderfully eccentric first Professor of Geology at Oxford in 1818, when a readership was created especially for him. Some consider him to have been the world's first palaeontologist.[33]

Although the idea of an over-large, mysterious and fantastical creature that had once lived in unknown aeons past was almost impossible to grasp mentally, to energetic Buckland, with his infectious enthusiasm, it presented no

problem. As he examined the fossilised bones of a complete specimen, millions of years fell away and the life of
the individual creature became immediate to him:

> When we see the body of an Ichthyosaurus still con-
> taining the food it had eaten just before its death; and
> its ribs still surrounding the remains of fishes that were
> swallowed ten thousand, or more than ten times ten
> thousand years ago; all these vast intervals seem anni-
> hilated; time altogether disappears; and we are brought
> into as immediate contact with events of immeasurably
> distant periods, as with the affairs of yesterday.[34]

Mary's *Ichthyosaurus*, as it was named in 1817, by
which time an even more fully complete specimen had
been assembled, was purchased for £23 (today about
£500) by the Lord of the Manor, Henry Hoste Henley, who
lived less than a mile away at Colway Manor in Lyme.
Mary is referred to as the person who paid workmen to
help remove it and who sold it to Henley, but this sounds
like part of the legend that grew up around her. She was
very young to have been in charge of such an important
transaction with the man who owned much of Lyme.
Widow Molly, who ran the business at this point, no
doubt would have been the negotiator. Molly, later
tantalisingly described as 'quite an original' herself,[35] was
to remain involved in what, after about 1825, became her
daughter's Fossil Depot until Molly's own death.

Henley was a collector for a private museum, and so
sold the specimen on to William Bullock's Museum of
Natural Curiosities, or 'London Museum', at 22 Piccadilly,
where it was exhibited in the newly built Egyptian Hall.
The ichthyosaur was one of a succession of imaginative

exhibitions organised by Bullock over the years, at various locations. The exhibits ranged from curiosities brought back by Captain Cook and memorabilia of Napoleon, to wonderful treasures from Mexico. Bullock, a member of the Linnaean and other learned societies, was noted for being the first to arrange a museum collection scientifically and systematically, using imaginative backcloths. That collection was dispersed by auction in 1819; the specimen became incomplete, and now only the skull resides in London's Natural History Museum.

The arrival of the specimen in London and attendant publicity coincided with the growing interest in the embryonic science of geology, as fossils of extraordinary creatures from the depths of time emerged to increasing publicity. It was widely but incorrectly believed that this specimen was the first to be found, and it was fully described in the *Transactions of the Geological Society of London* in 1814.[36]

But why were there fossils at Lyme Regis? Lyme is on the south coast of England, facing the sea from between two hills. In the sediments of the cliffs between Lyme and Charmouth are fossils from the lower Jurassic Period. It is one of the richest such areas in Britain and therefore has drawn collectors from all over the country and beyond.

During most of the Jurassic Period the Lyme area lay under a vast shallow sea that burgeoned with a prolific variety of life, providing a rich source of food for the larger carnivorous reptiles like the ichthyosaur. The Jurassic Period falls between the Cretaceous and the Triassic Periods. These three periods make up the Mesozoic Era, which runs from 250 to 65 million years ago, while 'the age of the dinosaurs' dates from about 230 to 65 million years ago.

BLUE LIAS

The appealing named Blue Lias comprises the sequence of rocks in which many of the entombed fossils at Lyme occur: 'Blue' because in some lights it appears to be blueish in colour, due to the blue-grey hue of the limestone, while 'Lias' may come from the Gaelic word for flat stone, or, more likely, from the Dorset quarrymen's pronunciation of the word 'layers'. The Lias is composed of layers, alternatiing in various degrees, of marls, clays and limestone. The sequence was laid down 200 to 197 million years ago in what was a seabed.

However, the friable Blue Lias was, and is, constantly under attack by Nature, from the outside by lively seas and weather, and from the inside by fresh springs forcing a way through. Added to this were the actions of man over several centuries; the Lias was useful in making stucco (waterproof cement), thus much of the Lias went to London for Regency building. Some of the older buildings in Lyme itself were built of the handy Lias, but this is now hidden from sight – because of its crumbling nature the material was often rendered to protect it from the elements.

In the Lias, all this activity resulted in fossils becoming partly or totally exposed, as they migrated to the surface and were then washed clear by the turbulent seas, new specimens often emerging after a storm in a never-ending process. But a violent sea could just as easily completely destroy and wash away a valuable fossil, and it sometimes took days, even years, in one case ten years, to excavate a large specimen.

Fossil reptiles, at the top of the food chain, are found in the Lias, including the large ichthyosaurs and the plesiosaurs, as are an abundance of other fossilised creatures, including ammonites ranging in size from tiny to massive. While Lyme has considerable deposits of Lias on all sides, to the west of the town, the Undercliff, this sequence is a staggering 400 metres deep. Within the Lias are twenty-one ammonite zones.

With great good fortune for the Annings, Lieutenant-Colonel Thomas James Birch (later Bosvile) (1768–1829) arrived on the scene, in 1818. This well-to-do fossil collector, a retired officer in the Life Guards, was touring the West Country, spending his half-pay pension in the search for fossils.[37] He went to Lyme frequently, and bought many fine specimens for his collection from the Annings. Birch visited Gideon Algernon Mantell (1790–1852) to view his collection, and the latter described Birch as a 'very agreeable and intelligent man'.[38] Mantell was a geologist and palaeontologist from Sussex, who was soon to become another star in the firmament of the new science.

In September of the same year at Lyme Birch found the most complete 'crocodile' (as the ichthyosaur was believed to be at the time and for most of the next ten years) to date. From this specimen the anatomy of the creature was first fully understood.

Birch would soon prove to be the Annings' saviour, for one day in 1819 he arrived at their shop to find them in dire straits and trying to sell off their furniture to pay the rent because they had not found one good fossil for nearly twelve months. Parish support had ceased some time before. Knowing the great contribution that the Anning family had already made, Birch could not ignore the situation.

Birch told Mantell what happened next, in his letter of March 1820:

I have not forgotten my promise to select for you some fine things from the blue lias – I cannot however, perform it yet as I have great occasion for every individual specimen I can muster. The fact is I am going to sell my collection for the benefit of the poor woman [Molly] and her son [Joseph] and daughter [Mary] at Lyme who have in truth found almost *all* the fine things, which have been submitted to scientific investigation . . . I may never again possess what I am about to part with; yet in doing it I shall have the satisfaction of knowing that the money will be well applied, the sale is to be at Bullock's in Piccadilly the middle of April.[39]

The title of the catalogue outlined what was to be sold: 'a small but very fine collection of organised fossils, from the Blue Lias Formation, At Lyme and Charmouth, in Dorsetshire, consisting principally of Bones, Illustrating the Osteology of the Icthio-Saurus, or Proteo-Saurus, and of specimens of the Zoophyte, called Pentacrinite, the genuine property of Colonel Birch, Collected at considerable expense, which will be sold by auction by Mr Bullock, at his Egyptian Hall in Piccadilly, on Monday, the 15th day of May, 1820.' Typical entries among the 102 items for sale were 'Fragment of a fish, the reverse side of the scales beautifully seen', 'Part of the foot of a prodigiously large animal . . .', concluding with 'This skeleton presents a most interesting illustration of the osteology of the Icthio-Saurus, or Proteo-Saurus . . .'.

The copy of Bullock's Sale Catalogue, now in the Earth Sciences Library, Natural History Museum, is inscribed on the cover 'This Catalogue belongs to the Fossil Shop, Lyme' and on the title page 'Joseph Anning, May 12, 1820, Lyme'.

The catalogue stated that the three-day sale would start precisely at one o'clock. Buyers came from as far away as Vienna and Paris, for the great Baron Georges Cuvier (1769–1832), the foremost anatomist and naturalist in

Europe, wanted to obtain several specimens. He suc-
ceeded in getting a selection including a femur and the
head of an ichthyosaur. Cuvier described this creature as
having 'the snout of a dolphin, the teeth of a crocodile,
the head and sternum of a lizard, the extremities of a
cetacea (being however, four in number), and the vertebrae of fish'.[40] The Royal College of Surgeons acquired
the Anning/Birch ichthyosaur for £100, along with other
specimens.[41] The sale brought in more than £400 (today
about £11,500).

What was the relationship between Mary, aged 21, and
Birch, aged 52? His concern and dramatic financial assistance naturally sparked rumours. In any story of a single
young woman, the question usually arises regarding
possible romances. A hint of gossip came from George
Cumberland (1754–1848), a painter in watercolours, who
wrote seven books about art, had a substantial collection
of fossils and was made an honorary member of the
Geological Society in 1810. On 24 August 1820 it was
Cumberland who recorded a somewhat indelicate
comment: 'Mrs Hanning [Anning] is the dealer at Lyme.
Col Birch is generally at Charmouth (they say *Miss
Anning attends him*).'[42] Tittle-tattle was to be expected in
this situation, at a time when women's lives were firmly
restricted and only the brave took outright action in
challenging or breaking out of those strictures, whatever
they may have thought, felt or controlled from behind the
scenes.

*　　*　　*

In May 1821 came another discovery by Mary. It was a
5-foot-long *Ichthyosaurus communis*. De La Beche

reported the find to the Keeper at the British Museum: 'the Annings . . . have found a very beautiful small skeleton. . . exceeding in preservation any yet found . . .'.[43] De La Beche, with much enthusiasm, immediately arranged for the British Museum to have first refusal at a price of £100. The museum, however, was not so enthusiastic. At that time, its intention was to save money, not to acquire the best example available; a much less perfect one was selected from the Annings for £50.[44]

More annoyingly, payment was exceedingly slow in coming. It had not arrived by September 1821, causing Molly Anning to take the unusual step of writing to the Keeper, who by then had apparently realised he had bought the wrong specimen and also felt that he had paid too much for it: 'As I am a widow woman and my chief dependence for supporting my family being by the sale of fossils, I hope you will not be offended by my writing to receive the money for the last fossil as I assure you, Sir, I stand much in need of it.'[45] This is the only surviving letter written by Mary's mother.

The finer specimen was acquired by the worthy Bristol Institution for the Advancement of Science, Literature and the Arts, which was first opened in 1823.[46] Fortunately, George Cumberland had recorded the name of the finder in a newspaper article: he noted that the specimen – so perfect that it 'sets at rest all further investigation' – was owed entirely to 'the persevering industry of a young female fossilist, of the name of Hanning [sic] of Lyme in Dorsetshire, and her dangerous employment'.[47]

Everything was now in place for Mary to make her mark on the developing world of geology. She was the right person, Lyme Regis was the right place and it was the right time.

Crucial to Mary's story was the transformation of Lyme from a mercantile town and port into a seaside resort attracting well-to-do visitors and collectors who could afford to buy fossils. The importance of Lyme had gradually been slipping, as the harbour was becoming too small to be used by ever larger vessels. But change was on the horizon. Before the turnpike road was constructed in 1758, 'a few invalids'[48] occasionally arrived to enjoy the benefit of the sea air. However, if they wanted to partake fully of the sea's benefit and dip in it – which was recommended – they had to undress on the beach. Homeowners near the sea therefore began to make a few rooms available for visitors.

Communications were improving: 'The arrival of the Post from London, is about five o'clock in the afternoon, and the western post at noon; the post-office was on Comb-street; the mail leaves for London at about nine in the morning.' The original horizontal and vertical posting slots are still there, which Molly and Mary must have used to post mail to the 'big-wigs'.[49] The upper vertical top slot may have been for those on horseback, or the belief may have been that this position made it more difficult to steal the post.

The roads continued to be improved, and this, with the introduction of carriages, transformed travel from being a kind of torture to be endured only when necessary into a comparatively pleasurable excursion.

Enduring on–off war with France, more on than off, had made travel to the Continent inadvisable, so pleasure was sought closer to home. Then there was the feeling of ebullience in undertaking a journey to areas of the British Isles previously considered to be impossibly inaccessible.

Also of importance, healthful benefits were to hand, backed up by several medical treatises. Dr Richard Russel, son of a London bookseller, had extolled the benefits of ocean water in the influential *A Dissertation on the Use of Sea Water in the Diseases of the Glands*, translated in 1750 from the Latin version he had written in 1749. Then he moved to Brighton himself. Other medical men were not slow in taking up the theme. Dr Crane of Weymouth recommended that bathing in the glacial sea was most beneficial in the morning and in cold weather, perhaps in January or February. There is no record of how many became ill or worse as a result of this treatment. However, the stern regime soon relaxed, more comforts were provided, and simply breathing the sea air was deemed to be beneficial. Perhaps there was no more risk than conventional treatments at the time, when quack doctors proliferated and effective medicines were lacking.

The pleasure-seekers had first ventured inland to famous spa towns like Bath (116 miles west of London), with its warm mineral springs, which pioneered the new trend. For some, however, this famed spa town, with its cohesion in perfect harmony of etiquette, dress, entertainment and architecture, became too successful, too crowded, and altogether too much.

The growing seaside mania came just in time to save Lyme from the creeping decay that was becoming evident. To many, the wildness and unpredictability of the British seaside became alluring, even symbolic of a libertine lifestyle. This coincided nicely with the rise of the Picturesque movement. A case was made for the appreciation of the unruliness of nature. The seaside, with its restless beauty and the sometimes tranquil, sometimes savage, sea was now regarded as the perfect destination.

However, when the diarist Fanny Burney visited Lyme briefly in August 1791 on a recuperative break, it was not to her taste; she found it 'dirty and fishy' and left quickly. And there was a limit to just how picturesque the view could be. In one Dorset village, the 'grandmother of a baronet' dreamily gazed out of her seaside window to see the unpleasing sight of a carcass of a horse strung up by one leg to the mast of a boat and flayed. It was conveniently located for the fishermen to hack off meat to use as bait.[50] The locals were edged towards adapting their ways to make themselves more acceptable to the paying visitors.

At the same time as those of certain classes were enjoying themselves, many others were struggling to survive. The agricultural labourers in Dorset were the most impoverished in the entire country. At times close to revolution, their ways of fighting back anonymously included hayrick burning, machine breaking and barn burning.

Even the royal family made their way to the seaside. George III chose the port of Weymouth, only 32 miles from Lyme Regis. His first visit was in 1789, the year of the French Revolution, when he had been suffering from the metabolic disorder porphyria for a year. The stay by the sea helped him recover from the initial attack, during which he appeared to be delusional and mad. Between 1789 and 1805, with attacks becoming more frequent and alarming, there would be fourteen follow-up visits to Weymouth, where he dutifully strolled by the sea and inhaled the bracing salty air.

The gulf between rich and poor continued to widen. At a time of desperate poverty and agitation for reform, the hedonistic life of the Prince of Wales (Prince Regent in 1810, and George IV in 1820) appeared in the most extreme contrast imaginable compared to the austerity of that of

Mary Anning and very many others. Always anxious to put some distance between himself and his far too sensible father, he chose Brighton, an adequate 120 miles east of Weymouth. The Prince's presence from the 1780s in what was originally a small Sussex village led to his creation of a fantastical folly there. The Marine Pavilion was a playful freak of architecture with its accumulative, derivative styles. It had an exterior that was turreted in the 'Hindoo' style, but, bewilderingly, was decorated without restraint on the inside in the 'Chinese' fashion. There the 322-pound Prince slept on sheets made of cooling silk satin till three in the afternoon, and enjoyed presiding over entertainments and banquets with as many as thirty-six entrées.

Encouraged by these two leading royal role models – one the essence of propriety, the other of raffishness – people rushed to enjoy the pleasures beside the English sea. The trend that had started with a trickle in the 1750s, by the late 1700s found the port of Lyme with its unique Cobb transformed. In its charming setting Lyme became a modestly fashionable seaside resort with assets that 'brought beauty and elegance in their train',[51] and, importantly, always boasting of sensible prices: 'Gentility without Ostentation' was the motto.[52]

The first pleasure pier was that of Ryde, Isle of Wight, built in 1814. Other resorts rapidly constructed a similar attraction – a 'repellant wooden pier' in historian Cyril Wanklyn's words – boasting of the thrill of actually walking out into the dangerous element, but with both feet on a firm base. Lyme, however, already had a unique 'pier' – the Cobb – its sinuous, elegant curves on two levels, with its slippery uneven upper surfaces sloping uncompromisingly towards the sea. 'After all,' said Wanklyn, 'there is only one Cobb.'[53]

THE COBB

What is the Cobb? It is a man-made harbour on a section of coast that had no natural haven. The Cobb was constructed in the reign of Edward I to the west of the old town, presumably where there was a shallower, harder base. Built by resourceful locals to safeguard Lyme from storms and create a safe harbour for fishing and trade, the protective, embracing arm curving out into the sea was a feat of engineering. 'Cobb' may come from the smooth cobbles from which it was first built, although historian and author John Fowles has suggested a meaning of 'rounded island'.

In the mid-seventeenth century Roger North left an account of this structure. He described it as 'a mole [a massive breakwater of stone] built into the sea about two furlongs [¼mile] from the town . . . There is not any one like it in the world: for . . . no stone that lies there was ever touched with a tool, or is bedded in any sort of cement; but all being pebbles of the sea, are piled up and hold by their bearings only, and the surge plays in and out, through the interstices of the stone in a wonderful manner.' If foundation stones were required, 'they searched them out upon the coast', finding the largest ones, and 'mounting them upon casks chained together with but one man mounted upon them, [who] with the help of a pole, conducts it to the place where it is to lie'. An iron restraining pin was struck out, the casks floated away and the stone fell into place.

Before 1800, when the tide was out, the goods landed were carried by horses east to the old town and the Customs House at Cobb Gate on the sands along 'The Road to Cobb'. The horses followed a mounted leader horse; when unloaded, they trudged back to do it again. When the tide flooded in, they were returned to their stables, and a boat was used.* The Cobb has always been a structure of national importance, for defence, for trade, and for supplying the Channel Islands during times of war with France.

* The Hon. Roger North, *The Life of the Right Honourable Francis North*, 3 vols, London, Henry Cobham, 1742, vol. 1, pp. 242–3.

Attempts were made to copy the facilities built for the pleasure of visitors at Bath, and so physical changes ensued in Lyme. Soon there were several bathing establishments with every convenience supplied, including private cubicles and warm water, seven newspapers and card tables. Bathing machines arrived, lodgings were improved, and more shops opened on Broad Street as well as hotels.

The Assembly Rooms were built as early as 1777, at the end of the Walk or Marine Parade on the square by Cobb Gate.[54] A description of them was given in the *Guide to Watering Places* of 1803: 'Lyme has a small Assembly Room, Card Room and Billiard Table all conveniently arranged under one roof.' Balls and card evenings were held in the Rooms, which became the focal point of a fashionable social life that might attract 200 people.

The Rooms, only yards from where Mary was born, provided an important social release for women of the right class. There it was acceptable for ladies to mingle socially with men to whom they were not related in a refined atmosphere that was set apart. There is a description of what the Rooms were like when Constance Hill visited in 1901: 'The ballroom is little changed since Miss Austen danced in it that September evening nearly one hundred [now 200] years ago . . . We visited the room by daylight, and felt almost as if it were afloat, or nothing but blue sea and sky was to be seen from its many windows. From the wide recessed window at the end, however, we got a glimpse of the sands and of the harbour and Cobb beyond.'[55]

At no point would Mary have been admitted socially to the Rooms. A woman 'in trade', as she was, was no lady. As late as 1880 the immensely wealthy Sir Henry Peek, the tea tycoon, was deeply shocked when refused membership in the Assembly Rooms because he, too, was 'in trade'.[56]

LYME AND LITERATURE

Lyme Regis punched above its weight in having links with literature. Not only have there been books, plays and poems referring to Mary Anning and her discoveries, but much else.

The visits of Jane Austen in Mary Anning's childhood led to the author setting scenes in *Persuasion* at Lyme (written 1815–16, published 1818); it may also have been the source of Mr Darcy's name in *Pride and Prejudice*. Did the author see the name of Captain D'Arcy, the engineer in charge of works on the Cobb in 1793, recorded on a bronze plaque on the Cobb?

In 1735 Henry Fielding, a handsome, unruly youth, tried to abduct his heiress cousin, Sarah Andrew, on her way to church, but her family repelled him. Sophia Western in Fielding's *Tom Jones* may have been based on Sarah.

There may have been a link with William Shakespeare. Famous Lymite Sir George Sommers (1554–1610), privateer and ad-venturer, sailed for Jamestown, Virginia, in 1609. His ship was caught in a hurricane and ran aground on Bermuda. Some survivors were reluctant to leave a land of plenty to confront the fury of the ocean. The castaways built two vessels and sailed on to Virginia.

One survivor was Silvester Jourdain (baptised Lyme 1565, died 1650?), who published his version of the event in 1610: *A Discovery of the Barmudas* [*sic*]. William Strachley wrote his account in 1610 (published 1625). Strachley knew Shakespeare, and his patron, Henry Wriothesley, Earl of Southampton, was by 1609 a member of the Virginia Company's Council, and must have heard about the adventures of the shipwrecked. The opening of *The Tempest* (Act I, Scene ii) and several sections of Act II seem to refer to Sommers's drama. Are the close similarities between *The Tempest* of 1611 and Sommers's story a coincidence? The great Shakespearian scholar Edmond Malone thought not.*

---

* Edmond Malone, *An Account of the Incidents from which the Title and Part of the Story of Shakespeare's Tempest were Derived, and its True Date Ascertained*, London, privately printed, 1808.

Bath is about 70 miles from Lyme, and during the peak
years of the seaside holiday, when the winter season in
Bath finished, Lyme's summer season began, lasting from
May to October. Little Lyme was 'humourously' described
as 'a kind of safety valve' for Bath, with which it was connected by a coach service. This led to the arrival of Lyme's
most distinguished and long-remembered visitor: Jane
Austen (1775–1817). When her father gave up his
Hampshire living, the family moved to Bath. They drove
to Lyme for a short break in 1803 (September and
November) and 1804, when Jane was 28 and 29. Through
her we have a rare view of what the town was like when
Mary Anning was a child of only 4 and 5. In some surviving letters and in her novel *Persuasion* Jane Austen
famously described Lyme and so immortalised the town
and the Cobb forever:

> The walk to the Cobb, skirting around the pleasant little
> bay, which, in the season, is animated with bathing
> machines and company; the Cobb itself, its old wonders
> and new improvements, with the very beautiful line of
> cliffs, stretching out to the east of the town, are what the
> stranger's eye will seek, and a very strange stranger it
> must be who does not see charms in the immediate
> environs of Lyme, to make him wish to know it better.[57]

In *Persuasion*, Louisa Musgrave famously fell on the steps
of the slippery Upper Cobb, landing 'lifeless' on the Lower
Cobb to be taken into the arms of Captain Wentworth.[58]

During their stay in September 1804, Jane noted that
'the servants make no difficulties' but 'nothing can exceed
the general Dirtiness of the House & furniture, & all its
Inhabitants'. However, the weather was 'just what we

could wish'. And it remained fine for their return visit that November – Lyme was famous for its pleasing, temperate weather. She went for walks, read, attended the dances in the Assembly Rooms and, using a bathing machine, bathed in the sea ('delightful', but tiring).

There was even an encounter with the Annings. In a letter dated 14 September 1804, Jane Austen mentions the 'Broken Lid' of a box. Richard Anning was summoned to examine it, and costed the repair at five shillings (25 pence, about £7.50 in today's money). Always careful with money, she objected to the price, saying that it was more than all the furniture in the (rented) room was worth, and referred the matter to the landlord.[59] Back in Richard's dust-laden shop, 5-year-old Mary Anning was soon playing round about the feet of her disappointed father once more.

Miss Austen was not yet famous. Although she had written most of her novels by the time of her visits to Lyme, they were thus far unpublished. But in Mary Anning's lifetime – by 1827 – canny Lyme landladies would be determinedly asserting that their cottages had been inhabited by the author herself or by one or all of her famous characters.

Improvements to the town were always required, and the *Dorset Directory* of 1823 noted plans in hand to make it equal to 'the most fashionable places in the west of England'. These included better paving, 'watching' and lighting the town, and a new road. Readers were assured of one of Lyme's plus points: 'the objects proposed in visiting a sea-bathing place may here be secured in a manner more compatible with the rigid rules of economy . . . lodging and board being exceedingly moderate.' The grocers, inns, milliners and so on are listed, and under 'miscellaneous', 'Anning, Joseph, furniture broker, Bridge-street' and 'Anning, Mary, curiosity dealer, Bridge-st' are to be seen.

Mary had become a professional hunter of fossils that she could sell, and serpent-stones or ammonites were good sellers to these visitors. One such visitor recorded that Molly and Mary had a 'tiny, old curiosity shop close to the beach . . . the most remarkable petrifactions and fossil remains were exhibited in the window'. On going inside one could see that the shop and adjoining room were 'crammed with ammonites, heads of crocodiles' and boxes of shells.[60]

Fossil shells and 'pretty little boxes of shells or taste-fully arranged bunches of seaweed'[61] were also among Mary's wares, and she included shells in the collection she assembled later for her little friend Miss Bell. There is a charming poem regarding seaweed, written by a long-time lady resident of Lyme Regis:

Oh call us not weeds, but flowers of the sea,
For lovely, and gay, and bright-tinted are we;
Our blush is as deep as the rose of thy bowers,
Then call us not weeds, – we are ocean's gay flow'rs,
Not nurs'd like the plants of the summer parterre,
Whose gales are but sighs of an evening air;
Our exquisite, fragile, and delicate forms
Are the prey of the ocean when vex'd with his storms.

This poem was written in an album under a 'tasteful' bouquet of dried seaweed.[62] Delightful as they are, the words conjure up a place that was not of Mary's world. A lady – identified as Miss Elizabeth Aveline, whom Mary no doubt knew – who could write such fine words was accustomed to pressing flowers at her tea table and making floral arrangements for the annual Flower Show held in the Assembly Rooms, with no worries about the daily looming threat of danger from the shifting cliffs and rolling seas. Mary's main interest in seaweed perhaps extended to *not* seeing it; she was delighted when storms scoured it off the foreshore, where it concealed the fossils she was after. No one could 'read' the interface where sea, land and water meet better than Mary.

Shells were a perfect memento of a visit to the seaside: small, beautiful, inexpensive, perhaps free. And in the great metropolis of London shell shops were popular, and had to be kept supplied. The Victorians enjoyed collecting, being out in the sea air searching for shells, obsessively building up their collections, or displaying their good taste by making elaborate pictures of shells and decorating mirror frames, bowls and trinket boxes with these often exotic beauties.

The sea is forever linked to shells, and shells are forever linked to Mary Anning because of the association of her name with a children's tongue-twister written by Terry Sullivan in 1908:

> She sells sea-shells on the sea-shore,
> The shells she sells are sea-shells, I'm sure,
> For if she sells sea-shells on the sea-shore,
> Then I'm sure she sells sea-shore shells.

The verse conjures up a soft and appealing image of what was a hard and dangerous life.

There was nothing remotely threatening about shells or seaweed. But fossils were an entirely different matter. Fossils may have once been considered to be medicinal,

SHELLS AND OIL

In the year 1833 a notable shop selling 'curiositoes' and shells was located in London's Houndsditch, an area by the moat that followed the City wall. It was a place of merchants, shopkeepers and used clothing shops. Here was merchant Marcus Samuel's Shell Shop. He found dealing in exotic oriental shells a lucrative line to add to his antiques and bric-a-brac, as the Victorian obsession with collecting and displaying shells took hold.

Samuel and his sons rapidly expanded the trade based on the Shell Shop into an import–export business with the Far East, for, when collecting seashells around the Caspian Sea in 1892, Marcus Samuel's son, also Marcus, saw the potential in exporting lamp oil from that area, the initial link with fuel. This business was to evolve into the oil giant Shell.

The first oil tanker in the world was commissioned by the Samuels in 1892 to ship kerosene to the Far East. The 5,010-ton vessel was named *Murex*, after a suitably exotic shell. The *Murex* shell belongs to the Municidae family, in which there are hundreds of species, most of them exquisite, like a conch shell, curving with whorls of frond-like spines, and displaying delicate colours.

The link with shells was maintained, and by 1907 the Samuels owned a fleet of oil tankers, each named after a different seashell, as they still are today. The first symbol in 1900 had been the less-than-memorable shell of a mussel. From 1904 the soon-to-be-famous scallop shell was introduced as the emblem.

Royal Dutch Shell, founded in 1907 and now trading in one-tenth of the world's oil and natural gas, carries as its logo a fluted shell, based on that of the giant sea scallop, *Pecten maximus* – to be seen almost everywhere worldwide.

The giant sea scallop shell has been used as an image by artists for centuries, since the time of the Greeks and Romans. This shell was also the symbol used by medieval Christians on pilgrimages.

magical, or even sinners turned into stone. But most seriously, fossils – especially those of the monsters that Mary Anning and others were beginning to find – challenged the account in Genesis of how the world was created. In short, fossils were dangerous. No one knew what they were or how they had got deep in the rock where they were found. Their existence provoked increasingly challenging questions that to many were disturbing and impossible to answer. Religious controversy was being deeply and irrevocably stirred.

2

Worlds within Worlds

The world was created precisely at 8 p.m. on Saturday
23 October 4004 BC.
(Scheme of biblical chronology calculated by Primate
of All Ireland, James Ussher, 1581–1656)

'Lord pity the arse that's clagged to a head that will hunt
stones,' wrote early geologist James Hutton (1726–97).
After hours in the saddle on an arduous journey through
North Wales in search of fossils in 1774, it was a reason-
able remark for a normally vigorous man to make.[1] In
short, geological research in the field and fossiling were
not for those weak in body or character.

Years later Mary Anning must have had similar sym-
pathy for her own body, when her head and heart – as
well as the prospect of no tallow candles to light a room
with a cold hearth – drove her to search for fossils she
could sell. Much of the time she was physically struggling against the unstable cliffs, the freezing rain and the

menacing sea. Mary became strong and tough. Not only did she survive, but also, most unusually, she became a female entrepreneur.

Mary had been born into an increasingly uneasy world where the structure of faith seemed to be under attack, as knowledge regarding the origin of the earth broadened to a degree that would have been unthinkable only a short time before. Many found comfort in the doctrines of the established Churches, but to some these no longer appeared so sound.

Five years before Mary Anning was born, Scotsman and gentleman farmer James Hutton, inquisitive, inventive and clever in several domains, was in the forefront of geological investigation and the inevitable conclusions to which that led. The prevalent belief system was seriously challenged when he published first a paper in 1785, followed by his *Theory of the Earth* (vols 1 and 2, 1795, vol. 3, 1899). He recounted his concept that 'the past history of our globe must be explained by what can be seen to be happening now'. These deductions were based on his experience on the land as a farmer on the Borders, years of intensive investigation, as well as extraordinarily detailed research and the collection of specimens.

As the eighteenth, then nineteenth, centuries progressed, industrialisation took hold, slowly at first, but as the population expanded, the need for transport of goods and people grew rapidly. With massive amounts of earth and rock being excavated all over the country, the strata of rock could be studied and fossils were revealed. Roadbuilding was led by Telford and McAdam. Water transport via canals, necessary for industrial growth, made the late 1700s 'The Age of Canals'; James Hutton was involved with the building of the Forth and Clyde Canal

– personal investment in it made him wealthy. In the transport network, railways were expanding, and quarrying for building materials such as limestone grew. Added to this was the increasingly important mining industry.

The result of this extensive digging in so many regions was an embarrassingly large increase in the number of important fossil finds, each adding to the complex puzzle of how life on earth had evolved, or confirming an aspect of it. And more and more questions were being asked.

Geologists were, of course, intimately connected with mining. In England, from De La Beche's knowledge of the mining communities came his idea that the nation should compile a geological map of the United Kingdom, practically and economically aimed at developing the mining industries; usefully, this also enabled him to collect and preserve specimens of rock (and led to the foundation of the School of Mines). The Government appointed him to develop what later became the Geological Survey of Great Britain, officially recognised in 1835.

Hutton's interest in geology had developed in the 1750s. During the next three decades he made extensive geological researches on journeys through much of England, Wales and Scotland. At the outset he observed – and he was not the first to do so – that the majority of rocks on the surface of the earth are formed from the debris of former rocks and that the earth's surface is gradually being destroyed by erosion. It was apparent to him that the covering of the earth was continually being recycled in never-ending processes.

Another central tenet of Hutton's theory was that immense forces of pressure and heat created in the earth over almost inconceivably long periods of time had thrown up seabeds (bearing fossilised shells and marine life) to become mountaintops. These were then folded, the tops eroded, and they were cast down again, sending mountains to the bottom of oceans.

Hutton was, however, the first to perceive the connection between these phenomena, arguing that the sediments produced by erosion must be consolidated on the seabed and then uplifted to form land.[2] The heat that was an agent of consolidation and uplift also generated in the interior of the earth hot fluids from which crystalline rocks originated. Hutton went further, claiming that erosion, uplift and igneous activity were continuous processes that had always, and would always, operate in the same way and thus that the surface of the earth was continually being recycled. Hutton famously stated, 'we find no vestige of a beginning – no prospect of an end'.[3] This was the essential point.

At the time of the publication of his first two volumes of *Theory of the Earth* in 1795, Hutton was still uncertain whether or not granite was igneous (formed from magma that has cooled and solidified). That summer he made the first of his three famous excursions to prove this point. In the bed of the River Tilt, in Perthshire, he found the junction between the granite of the Cairngorms and the 'marble' (schists and limestones) of the mountains to the south. Branching veins of granite had penetrated deep into the 'marble', which could have happened only if the granite were the younger, once hot molten rock which had been forced into the 'marble' *from beneath* by immense pressure and the subterranean heat.

This provoked fierce argument, for it meant that the state of the earth's surface could *not* be the result of the force of flood waters – or The Flood. Hutton and others were studying the rocks themselves, not the Bible, to determine the course of events. And did Genesis 6:4 not record that 'There were giants in the earth in those days', before the Noachian Flood?

In the ensuing arguments and conflicting theories, the experts broke into several camps. On the Continent Abraham Werner (1750–1817), a German geologist and mineralogist, was famed for his teaching on mineralogy at the Mining Academy of Freiburg. He brought order to the classification of rocks, grading the earth's surface into four distinct types, always found in the same succession. Believing that all rocks were formed through the action of water, he concluded that the oldest rock had arisen from a Universal Ocean, and was more than a million years old. Hutton, of course, did not accept this.

When William Buckland, frequent traveller that he was, visited Weimar in 1816, he met Goethe and Werner: the latter was most hospitable to the eccentric Englishman, but studiously avoided any mention of the Creation and geology.[4] Werner's theory in itself was shocking enough to the Creationists, who held the belief that the landscape of the earth had been formed by the Flood, and that all life dated from then. There was no means of establishing the dates for the rocks and fossils, so people turned, as they had always done, to the Bible.

The span of Jean André De Luc's life touched those of both James Hutton and Mary Anning. Hutton had recognised that De Luc was a formidable opponent. Raised a pious Calvinist, De Luc was a staunch defender of the account in Genesis. He attempted to reconcile the biblical account with geology, concluding that the six days represented six epochs, which ended with the Flood; at that point the earth's surface attained a permanent form as a result of a series of earlier catastrophes.

COLLIDING BELIEFS

The Creationists followed the account given in Genesis strictly, but at this time of uncertainty various sects arose, each claiming to have the answer, each vying for position. Abraham Werner's followers belonged to the geological sect called Neptunists, who held that the action of a primordial sea had shaped the earth's surface. The Vulcanists, on the other hand, held that the world had been formed by the action of volcanoes. The Catastrophists claimed that the earth arose after a series of catastrophic floods, while the Uniformitarians – accepting Hutton's theories – stated that the earth had experienced slow change over immense periods of time.

Natural and divine philosophy, though separate, were not incompatible, Hutton had argued. But his conclusions were at odds with a belief system dominated by the Old Testament version of Creation in which the world was created literally in six days, and man was God's greatest, most special creation. Every living creature dated from the time of the Flood and still existed somewhere, the belief being that God would not permit any animate thing he had created to become extinct. While this was just about possible to accept for smaller creatures, it was impossible to make sense of it as far as the 'giant lizards' being discovered were concerned. The fossilised fauna were obviously earlier, prehistoric forms of life, but there was no prehistory in the Bible.

Hutton was a theist, consistently asserting that a beneficent deity had designed all the operations of the earth for the ultimate benefit of man, but his theories could not help but offend many Christians by implicitly rejecting the biblical account of the Creation and Bishop Ussher's uncompromising and carefully calculated chronology for the age of the earth.

However, the arguments and beliefs of each group were often fluid, and supporters came and went. In Mary Anning's later years the views of the Uniformitarians were the most widely accepted.

The sequence of events that Hutton outlined could only have taken place over an enormous period of time. He had advanced this explanation with great eloquence, making an indelible impression on those attending his lectures who 'seemed to grow giddy looking so far into the abyss of time'.[5]

It was not until after Hutton's death that his suppositions were verified by experiment. His theory of the earth was studied on the Continent and in America but attracted few supporters. Fifty years were to pass before his ideas would be accepted.

In France, the Revolution of 1789 broke the stranglehold of the Roman Catholic Church. There, Cuvier held the view that the earth had suffered and continued to suffer from a series of catastrophes, when one group of creatures was wiped out to be replaced by another. He was totally opposed to the theory of evolution, believing that each creature had been made by God for a specific purpose. However, Cuvier wisely kept religion and science separate.

Mary Anning would have been aware of Cuvier at the Muséum National d'Histoire Naturelle in Paris, at least from 1820, when he purchased a selection of fossils (that she had found originally) at Colonel Birch's sale. Cuvier was considered to be the leading European intellect of his age, and he would reappear in Mary's story.

However, in England, the Church of England retained its power, continuing to be a straitjacket on intellectual progress. Fossils and the questions they raised caused alarm among clerics, who saw faith in Genesis being directly confronted. George Cumberland reflected the general thinking in 1815: 'We want no better guide than Moses.'[6]

It was John Playfair (1748–1819), a Scottish mathe-
matician and geologist, who promoted Hutton's theory of
the earth most effectively in *Illustrations of the Huttonian
Theory* (1802) using the river valleys of southern England
as examples. Largely because of this book, Hutton's ideas
were taken up by the following generation of geologists,
most notably by Charles Lyell (1797–1875), who wrote
Principles of Geology (three volumes, 1830–3), which
Darwin took with him on his voyage. From the mid-
nineteenth century onwards Lyell's suppositions were so
widely accepted that Hutton's name was sometimes
forgotten. For many Christians, it was too shocking to
contemplate the thought that the calculations of
chronology in the Bible and by God's representatives on
earth were wrong – not by thousands, but by millions of
years. One result of all this speculation was that the early
fossil hunters were looked on with suspicion, and
'almost universally viewed in a bad light as infidels and
perverters of the Scripture'.[7] But there was a great deal in
the controversial 'new geology' to appeal to the
Victorians: making detailed observations, obsessively
collecting rocks and fossils, forming collections, and
relishing the prestige of donating specimens to the new
museums that were being established.

* * *

How did all this affect Lyme and Mary Anning? In Lyme,
fossil collecting became an increasingly popular way to
earn extra money, thereby giving Mary unwelcome
competition. Even John Gleed, Mary's pastor from 1818
to 1828 at the Independents' Chapel, collected fossils to
sell and help improve his income. And when Gideon

Mantell was in Lyme in June 1832, he visited Mary's shop and then sought out a fisherwoman who had some fossils, purchasing from her the vertebral column of a *Plesiosaurus dolichodeirus*.[8]

The extraordinary fossils that Mary and others were finding led to awkward questions at a time of social unrest when the desire was strong to cling to the certainty of the old religious beliefs that had sustained generation after generation. For most, however, the Noachian theory of Creation remained firmly in place until the 1830s.

There are some clues as to how Mary was able to reconcile these two views. She was a devout chapelgoer, yet all around her was evidence that evolution had taken place. Hints as to her thinking are revealed in the story of her friendship with a small boy. The Revd Francis John Rawlins, as a child, became friends with Mary Anning in 1833, when his family, who had various connections with the area, went to Lyme for a seaside holiday. His father then sent Frank, as he was called, to Roberts's Crewkerne School in Lyme.

The serious 6-year-old 'was attracted by a fine display of fossils in the window of a small curiosity shop, the owner of which was a remarkable woman'.[9] The shop contained mainly vertebrate fossils, although a few other items would have been on display. The woman, whose kindness to children was so often noted, responded to the boy's keen interest and encouraged him to learn about fossils, even, rather charmingly, helping him to make labels for his collection; on each one he wrote the place and the depth at which it had been found.[10]

When still very young, Frank tried to understand how the fossils found at various undisturbed depths could be explained in accordance with the account of Creation given in Genesis. He asked his father, the Revd H.W. Rawlins, to explain it. As an evangelical pastor, Revd Rawlins interpreted the Bible literally, replying that all such monsters must have perished in the Flood and all existing creatures dated from after the Flood. As Frank's brothers and sisters listened quietly, the boy asked how it was possible that the creatures being found could have burrowed their way deep into the rock and died there. His father would hear none of it, and accused him of being childish. He was a child, but, even then, felt that someone must have made a mistake. He kept his thoughts to himself, but later as an undergraduate he secretly read Robert Chambers's (1802–71) *Vestiges of the Natural History of Creation* (1844),[11] which aroused controversy, dismay, passion and fury. In it Chambers argued that the universe had not been created by a single act of God, but had developed and was guided by scientific principles. The book was published anonymously, and – written in a popular style – went through four editions in seven months.

What did Mary think? She was devoted to the Congregationalist chapel, often giving what money she could spare to it, but at the same time she realised that there had been a progression of living things. In June 1844 she wrote a letter to her friend Mrs Dorothea Solly, daughter of the rector at Spettisbury and Charlton, Thomas Rackett, a noted conchologist. At this time Mary was nearing the end of her life and she was contemplative. As always, she did not use punctuation and had her own way of spelling. She wrote: 'I can only remark on it generally as truly believing from what little I have seen of the fossil World and Natural History I think the connection or analogy between the Creatures of the former and present World excepting as to size much greater than is generally supposed . . .'.[12] It would seem that, like Hutton, she was a theist – that is, she believed in God as the creator of all for the benefit of mankind, but came to accept evolution as part of God's plan.

In an undated essay of about 500 words copied out by Mary (in her handwriting, but unsigned), in the Richard Owen correspondence at the Natural History Museum, she noted:

> The family of ammonites, for instance, contains more than two hundred fossil species according to many authors, and it does not seem possible to reduce this estimate above one half; yet of all those not one is known recent, and the only recent species of the whole genus is a very minute shell; yet the fossil species sometimes measures three feet in diameter is it probable that a genus so numerous, and having species of such large size, can have been overlooked . . . the same remarks will apply to the belemnites of which no recent species is known. The land quadrapids [*sic*] found in some of the most recent strata, and many of these even mingled in the diluvial detritus still exist in the same countries.

Mary continued, 'M. Cuvier has shown at large the little probability there is that any of them exist in an unknown condition.' She emphasised, 'it must be carefully remembered that an accurate and vigorous knowledge of zoology is requisite in anyone who ventures to discuss the subject; a superficial acquaintance with it can only lead to confusion and error.' She noted the great differences among the fossil groups from one to another and also those from the more contemporary creatures she had seen or read about. She thought it unlikely that any fossil faunas from the distant past were still living. Conversely, she notes how some invertebrates, such as the bivalve *Trigonia*, were known in fossil form long before they were found as living creatures.[13] The daughter of individualistic parents (her mother was Church of England), she was a woman with a mind open to new ideas at a time when most minds were closed.

It is tempting to think that shrewd Mary decided that this was one issue it would be most wise to sidestep and leave to scholarly gentlemen – with their time, money and prestige – to battle over. Meanwhile, she quietly carried on working, using her great knowledge to earn a living, with her faith to sustain her as it always had done and would continue to do.

*   *   *

In fact, many of the prominent pioneers of the new geology and the fossil hunters – while seeming to threaten the biblical account of the Creation – were themselves churchmen and/or sons of churchmen. Mary knew them well, including William Buckland (ordained in 1809 at the Chapel Royal, St James's), Revd William Daniel Conybeare (1787–1857) (ordained as a deacon in 1813), and Adam Sedgwick (1785–1873) (offered a living in 1832) at Cambridge, with whom she would later correspond and to whom she would sell fossils, some of which remain in the Sedgwick Museum of Earth Sciences, Cambridge. Added to their passion for fossils was a belief in the teachings of the Church. These men remained circumspect, bearing in mind that the Church was also their employer, in many cases the source of a very comfortable living, and had established or confirmed them in their own high social status. Church appointments, with their prestige, security and financial rewards, enabled these scholarly men to pursue their studies in comfort.

In the same year as he was ordained, William Buckland, amiable, of average height and good-looking, was made a Fellow of Corpus Christi College, Oxford, placing him firmly in a deeply conservative Church of England environment where almost all the professors were ordained ministers in the Established Church. However, he was well aware that religious beliefs were being shaken by the discovery of fossils. He himself simply could not accept that mammals had evolved from lizards, and sincerely sought ways to reconcile the Bible with the prehistoric discoveries increasingly emerging from the rocks. Buckland chose to interpret the assertion that the world had been created in six days, not in a literal sense but in an abstract way: a main geological period corresponded to a period of human history. The Flood had occurred but it had been the final disaster in a series of earlier cataclysms, from which the rocks and matter comprising the earth today are the result. This was somewhat similar to the theories of De Luc and Cuvier, and was sufficiently vague to allow Buckland to operate with sincerity and dedication in both fields.

At Oxford, Buckland, a Reader in Mineralogy since 1813, had quickly become famous as a lecturer. Through the brilliance of his entertaining lectures, given at 2 p.m. three times a week, he was immensely influential in making the new geology the queen of sciences. Charles Lyell – who became the leading geologist of the

mid-nineteenth century and an influence on Charles
Darwin – was one of his students, as were chemist and
botanist C.G.B. Daubeny and collector Sir Philip Egerton.
Buckland himself fell under the influence of William
Smith (1769–1839), the 'Father of English Geology'.
Smith was the first to recognise that strata were laid
down in the same sequence all over the country.

Buckland continued to attempt to reconcile the two
views of Creation by asserting that one does not look to
the Bible for 'scientific truth' but for God's revelation and
for instruction in 'the divine life'.[14] His colleague Philip
Nicholas Shuttleworth wrote:

> Some doubts were once expressed about the Flood;
> Buckland arose, and all was clear – as mud.[15]

George Brettingham Sowerby (1788–1854), a dealer in
shells and other natural history items, told a wonderful
story of when he was out fossiling with Buckland. The
latter, on finding a large ammonite with the central coil
missing, carried it home by putting it over his head and
galloping off on his trusty black mare.[16] Buckland must
have been enormously strong with an extraordinary sense
of balance to have done this.

Buckland's involvement in this controversial new field
of study – the 'noble subterranean science', as he called
it[17] – was causing ripples, even waves, in the tranquil
waters of Oxford. When he went on a tour of the Alps
and Italy, one of the Oxford dons retorted dismissively,
'Well, Buckland has gone to Italy; so, thank God, we shall
hear no more of *this geology*.'[18]

In 1818 William Buckland decided to rally his
impressive and influential contacts to campaign for a

Chair in Geology at Oxford. The Prince of Wales had been Regent since a seriously debilitating illness had taken hold of George III in 1810; His Royal Highness became involved in Buckland's campaign, and the Government found the money. In return it was confidently expected that Buckland would find the geological evidence to support Genesis.

The first Professor of Geology at Oxford, aged 40 in 1825, was still surviving on £200 per year from two Readerships, when he obtained a highly desirable prize: a canonry at Christ Church, Oxford. This meant more than adequate accommodation, in his case an extensive set of rooms at Christ Church off the main quadrangle for him and his new wife, also a Mary, £1,000 per year (almost £30,000 today) and, most usefully, no requirements of duty or residence.

Buckland had an infectious enthusiasm and was immensely likeable; people responded well to him. His idiosyncrasies were more than tolerated; they were enjoyed and relished, and the stories retold. Roderick Impey Murchison (1792–1871), an undergraduate, eventually to become an eminent geologist, went on his first field trip with Buckland. He left a record of the sight that greeted him one day on entering Buckland's Oxford apartments: 'On repairing from the Star Inn to Buckland's domicile, I can never forget the scene that awaited me . . . Having climbed up a narrow staircase . . . I entered a long corridor-like room filled with rocks, and bones in dire confusion. In a sort of sanctum at the end was my friend in his black gown, looking like a necromancer, sitting on a rickety chair covered with some fossils, clearing out a fossil bone from the matrix.'[19] In addition to the fossils and stuffed animals in the hall, candles set in the vertebrae of *Ichthyosauri* in the dining-room flickered over a scene further enlivened by cages of live snakes and green frogs. Added to this, some all too vigorous creatures were on the loose, including five guinea pigs and a jackal. When tutor Walter Stanhope visited, he noted, 'I took care to tuck my legs up on the sofa, for fear of a casual bite from a jackal that was wandering around the room.' This was a wise move, as munching noises soon emerged from under the couch and on investigation it was apparent that four of the five guinea pigs had been devoured.[20]

There was more. Dining chez Buckland was an adventure in itself because almost any animal might be served up, from toast of mice (not bad) to crocodile (not good), as he experimented with his eminent guests' taste buds.[21] He had a noble aim in this seemingly bizarre behaviour: genuinely concerned, he was researching additional food sources for the poor to eat. And there was no limit to his experimentation. His extravagant eccentricity transgressed into the surreal on one occasion when he was injudiciously handed a silver casket in which reposed the embalmed heart of Louis XIV. He instantly proclaimed that he had never eaten the heart of a king – and it disappeared for ever into Buckland's mouth![22] Most extraordinary of the inhabitants was a pet bear, Tiglath Pileser, who, decked out as a student, attended university functions and especially enjoyed wine parties.[23] This behaviour, as well as his sometimes coarse language, was, of course, beloved by the students:

> Here see the wrecks of beasts and fishes
> With broken saucers, cups and dishes;
> The Pre-Adamite system jumble,

With sub-lapsaurian bruises tumbled,
And post-Noachian bears and flounders
With heads of crocodiles and flounders;
Skins wanting bones, bones wanting skins
And various blocks to break your shins.
No place in this for cutting capers,
Midst jumbled stones and books, and papers,
Stuffed birds, portfolios, packing-cases,
And flounders fallen upon their faces . . .
The sage amidst the chaos stands,
Contemplative with laden hands,
This grasping tight his bread and butter
And that a flint, whilst he doth utter
Strange sentences that seem to say,
'I see it all as clear as day.'[24]

His wife, Mary Morland Buckland (1797–1857), was a geologist and gifted scientific illustrator in her own right. She was a match for her husband in intellect and interests, with her own entry in the 2005 *Dictionary of National Biography*. Before her marriage she had worked closely with no less an eminence than Cuvier and had illustrated some of his works. She supported her husband in every way, including making models of rare fossilised creatures out of leather and identifying all the specimens with labels, a laborious, never-ending job with which their nine children were expected to help. A visitor might, in this happy, turbulent household, be surprised when three children on the back of their pony charged through the front door, up the wide steps, past the stuffed creatures and the rocking horse in the hall, raced round the dining-room table and back out again. The eldest, Frank, would become a distinguished naturalist and pisciculturist.

Mary Anning had known Buckland from the time of her first discovery. Their working relationship spanned the years, and she must have been entertained by his colourful, relentlessly cheerful and unpredictable character. She corresponded with both William and Mary Buckland, the latter being someone with whom Mary could converse comfortably. She enjoyed the Buckland children and got to know them well, for when William first visited Lyme with his family, he 'took the children fossiling and made them acquainted with . . . Mary Anning'.[25] Mary ended a letter to Buckland of 21 December 1830 with affection: 'How is Mrs Buckland is she confined yet? And the Dear children how are they and are you coming down at Xmas.'[26] In another letter she charmingly sends them 'a mouthful of kisses'.

Buckland eventually took up the theme of glaciation to account for the arrangement of rocks on the surface of the earth. In 1836 his contribution to the series of books known as the Bridgewater Treatise was *Geology and Mineralology Considered with Reference to Natural Theology*. In it, he abandoned the concept of the Noachian Flood in favour of a theory of progression of life on earth. However, he never gave up on trying to reconcile the new geology with religion.

* * *

Revd Adam Sedgwick entered Mary's story later, around 1840. He was Professor of Geology at Cambridge for fifty-five years, and had himself undertaken geological research in the quest for the truth. One of the unmoneyed scholars, he gave lectures that always placed geology within natural theology, endorsing the moral and spiritual basis of Christianity. In 1844 he was deeply disturbed by the publication of the already mentioned *Vestiges of the Natural History of Creation* written anonymously by Chambers. Sedgwick's response was that, if true, 'religion is a lie; human law is a mass of folly; morality is moonshine; and man and woman are only better than beasts'.[27] Charles Darwin had been one of Sedgwick's students, and Sedgwick would have been much more seriously distressed by another book, published in 1859, three years after his death: Darwin's *Origin of Species by Means of Natural Selection.*

Others besides Mary Anning were, of course, making an increasing number of finds, challenging the Bible ever more strongly. One of these was Gideon Mantell (1790–1852), who wrote *The Fossils of the South Downs or Illustrations of the Geology of Sussex* in 1821. He would go on to concentrate his interest on fossil reptiles and fish. Mantell visited Lyme in June 1832: 'after a long and tedious drive, sweeping round the crest of the hill, we descended by a fearfully precipitous loose and dangerous road to Lyme . . . A more beautiful day never shone in England; and although the journey was to me rather fatiguing from my having risen so early, yet it was one of the most interesting and gratifying I had ever undertaken.'

Mantell left an unflattering (therefore interesting and probably true) description of Mary at that time: 'we sallied out in quest of Mary Anning, the geological Lioness of the place . . . She, the presiding Deity, a prim, pedantic vinegar looking, thin female, shrewd and rather satirical in her conversation. She had no good specimens by her, but I purchased a few of the usual Lias fossils.'[28] Perhaps he had been sarcastic to her and in return had been on the end of her ascerbic tongue.

Mantell's attractive, intelligent wife, yet another Mary (1795–1869), undertook the drawings for the first of his twelve books. In a story that may or may not be legend, in 1820 or 1821 Mary Mantell found a strange fossil tooth. This reputedly happened when she was taking a stroll to pass the time while her husband was seeing a patient. On his return Mrs Mantell presented it to her husband. Worn down and 1 inch long, it could hardly have caught the eye of a better-placed person. Her undeniably clever husband perhaps was reluctant to share the glory, or at the least did not clarify what became a rather vague story. Until then all primeval monsters were believed to have been carnivorous. But Mantell knew that this was something that had never been seen before: the tooth of a giant herbivorous reptile. However, when he showed it to members of the Geological Society, they dismissed this son of a shoemaker, who had not been able to attend grammar school or university because of his father's well-known political and religious views. Then Cuvier saw the tooth and discouragingly declared that it belonged to a rhinoceros.

Eventually, when presented with more evidence, Cuvier agreed with Mantell that this seemed to be a herbivorous reptile – that is, a previously unknown creature. Mantell then had the best possible feeling of satisfaction when he went to the next meeting of the Geological Society with the letter from Cuvier justifying his theory. He once again displayed to the Fellows what turned out to be the tooth of an *Iguanodon*. This time he was welcomed into their company.

One of the highlights of Gideon Mantell's career occurred in 1834 when he acquired an almost complete *Iguanodon*, so named because it had teeth and bones similar to an iguana. The stocky *Iguanodon* walked on its hind legs or on all fours, and grew up to 40 feet long. It was a herding dinosaur living 145 to 100 million years ago in the Early Cretaceous Period, which followed the Late Jurassic.

If it had been this difficult for Mantell, a male and a scholar although from a working-class family, to be accepted, how much more difficult would it have been for a woman, especially one with Mary Anning's background? Throughout his life, despite all his energy and intellect, Mantell suffered greatly from a lack of funds and, like Sedgwick, came from a poor home. However, most of the eminent gentlemen whom Mary knew were wealthy to some degree. Buckland came from a well-to-do family, and found himself a cosy and brilliantly successful sinecure at Christ Church. The death of Conybeare's father in 1815 left him well off for the rest of his life. Murchison was comfortably off, added to which his wife would become an heiress, and William Hawkins, the son of a farmer, somehow played the game – no one knew quite how, possibly by marrying a succession of wealthy wives – to give the impression and live as if he had ample funds.

The substantial income of Mary's old friend Henry De La Beche was based on a plantation in Jamaica, although this income was increasingly at risk as the campaign for the abolition of slavery gathered momentum.

Henry De La Beche (pronounced Beach), the son of an army officer, was born in London. The little De La Beche family was visiting Henry's father's recently inherited plantation in Jamaica in 1801 when De La Beche senior died. Thus, 5-year-old Henry, the only son, inherited the landed property and slaves. Back home in England, surviving shipwreck on the way, his mother remarried, but the second husband died. Henry and his twice-widowed mother finally settled in Lyme Regis with her (by then) third husband, Mr William H. Aveline, who himself had an interest in geology and often used to accompany the ladies fossiling. They lived in Broad Street, in a large house halfway up on the east side.[29]

De La Beche spent his youth in Lyme, where he clambered over the valleys and cliffs, indulging his love of geology. He was about the same age as Joseph Anning, and he soon met and worked with Mary, so beginning a life-long friendship. Conybeare later wrote of De La Beche's habit of exploring the cliffs around Lyme Regis with Mary Anning.[30] After a brief spell at military college, from which he was ejected for insubordination, he returned to Lyme. At this point he became friends with two men in Lyme who were interested in geology and fossils and went on geological expeditions: Dr Carpenter, who collected fossils, and meteorologist George Holland. This, along with the influence of the Annings, set his course for the future. Over the coming years, De La Beche and Mary searched for the fossils of reptiles, working as an excellent team – De La Beche had the education and Mary had the practical experience. He never failed to give her the full credit that she deserved.

In 1817 De La Beche joined the Geological Society of London, formed in 1807 by geologist George Bellas Greenough (1778–1855) – friend of Buckland and Conybeare – and incorporated by Royal Charter in 1825, reflecting the growing interest in this new science. Among the founding members were Buckland and Dr James Parkinson, who identified the condition known as Parkinson's Disease. Parkinson was a notable character who was a surgeon, social reformer, geologist and

SLAVERY, WHITE AND BLACK

Slavery comes into Lyme's story, and into the lives of several people closely connected with Mary Anning. For centuries mariners from the West Country, including at least twenty-five men from Lyme, had suffered being captured on sea or land by Barbary pirates from North Africa. In some cases ransom money was collected for their release, and at least one captive returned home to Lyme. This distressing danger was at its worst during the fifty-five-year reign of the mercurial and extraordinarily cruel Sultan of Morocco, Moulay Ismail (d. 1727). The options for those captured were converting to Islam, spending the rest of their short lives as slaves, or death.

At the same time as the inhabitants were suffering from this trade in European slaves, there was a much more odious and extensive trade in African slaves. The British were the main slave traders in countries bordering the Atlantic. Lyme itself was involved, as were many English ports. Over the centuries ships had docked at Lyme from exotic places like Guinea, the West Indies, Portugal and Spain, bearing commodities ranging from gold dust to coal, tobacco to tea, elephants' teeth to slaves, and salt cod from Newfoundland.

Mary would have seen the occasional dark face, perhaps that of a servant, especially in her younger years when the port was busier. Ships from Lyme were involved in the triangular trade: England to Africa to America to England. Home and European products were exported to Africa; from there slaves were taken to America and the colonies; the final stage brought back the much enjoyed sugar, tobacco and cotton.

The Atlantic slave trade had been abolished in 1807, but owning slaves remained legal. By spring 1833 Lyme, too, was preparing petitions to be sent to the House of Commons and the House of Lords supporting abolition, and in this year slavery was abolished throughout the British Empire. Full emancipation finally arrived in 1838. In the USA abolition came in 1865 after the Confederate defeat in the Civil War.

palaeontologist. In 1804 Parkinson had published *Organic Remains of a Former World*, which Mantell called the first attempt to give a familiar and scientific account of fossils. The Society initially had rooms in the Temple, then Covent Garden, and then larger quarters at Somerset House. Finally, the Society came to rest in Burlington House, Piccadilly, where it remains today. The Society was an important meeting place for geologists from all over the country at this time of great change. Its aims were, and remain, to promote knowledge concerning the history of the earth through lectures, publications and library services. Many of the geologists knew Mary Anning, and would have visited Lyme to purchase fossils from her – and to pick her brains.

De La Beche was directly involved in slavery, and found himself painfully pulled in two directions. By all accounts a good and honest man, with enlightened self-interest, he considered himself to be 'a well-wisher to the Slave population', whose own slaves, he claimed, were well cared for and contented. He sailed from Bristol to visit his 4,500-acre plantation in Jamaica for a year, from November 1823 to December 1824, becoming an expert on the geology of the island. He argued for the gradual introduction of reform, for which he received criticism. As an example, he ordered that overseers in the field were not to carry whips but that punishments were to be effected later.[31]

On his return to England De La Beche published *Notes on the Present Condition of the Negroes in Jamaica* (1825) and tried to defend the system. In letters he argued the case with Conybeare, an ardent abolitionist who believed that slavery was 'detestable and intolerable', claiming that the situation had been misrepresented and calling for a commission to be appointed. But he knew that the end was approaching.

By the early 1830s the flow of once considerable funds to De La Beche ceased, and it was clear to him that he would have to work for a living. In 1831, he wrote the *Geological Manual* – his most successful publication among many, with three English editions, as well as those in French, German and 'American' – and from 1832 he worked at adding geological information to the country's topographical maps, beginning with Devon. Thus, it fell to Henry De La Beche to pioneer the career of the professional geologist.

Mantell, sharp in his description of Mary, was equally harsh in a later reference to De La Beche, whom he described as once being 'a very rich West Indian proprietor [who] travelled a good deal, but from various causes has been much reduced in circumstances and is employed by the Government to colour geographically the ordnance maps'.[32]

The anti-slavery crusader William Wilberforce (1759–1833), who coincidentally had attended Buckland's lectures at Oxford, visited Lyme in 1804 for a long rest at the end of his abolitionist campaign. In fact he was in Lyme at the same time as Jane Austen, although there is no indication that they met.[33]

*   *   *

Some time in the 1830s there was a change in Mary's religious conviction because she and her brother switched to the Church of England. The exact reason is not known. Even back on 27 June 1799, when she was baptised in the plain Independents' Chapel, there had been a problem. A Charmouth pastor, the Revd John Crook, officiated because the regular pastor had been dismissed the year before; shockingly, it was discovered that he was an Arian, meaning that he did not accept the divinity of Christ.[34] Much later, when Mary was 18, the pastor James Wheaton died. The new pastor, Revd John Gleed, lived near the Annings and was interested in fossils. But after a time there was trouble in the church and many left. Sister and brother may have been disenchanted, too. Once highly regarded, the Independents' Chapel almost disbanded in the 1830s. Even the pastor from 1828 to 1838, Revd Ebenezer Smith, defected to the Established Church.[35]

Mary's mother, Molly, had been raised in the Church of England – some of Molly's children were baptised at the Coombe Street Chapel and some at St Michael's. Mary may have increasingly aligned herself with the Church of England and the Establishment. The conflict between the Creationists and those who could see that evolution had taken place would not go away. Mary, however, perhaps felt supported in her pioneering work by female companions, especially other fossil collectors. She was part of a small group of intelligent women resident in Lyme, and she formed friendships with some of the female visitors to Lyme, who sometimes stayed for lengthy periods.

Her perception was well noted. John Murray (1778–1843), the science writer, recalled meeting Mary Anning: 'I once gladly availed myself of a geological excursion and was not a little surprised at her geological tact and acumen. A single glance at the edge of a fossil peeping from the Blue Lias revealed to her the nature of the fossil and its name and character were instantly announced.'[36]

The wealth of other families was also based on the slave trade. One was the Pinney family, to become prominent in Lyme, and so important to Mary's story in leaving behind personal details in Anna Maria Pinney's journal.

Miss Pinney (b. *c.* 1813) was the daughter of J.F. Pinney, Esq., of Somerton Erleigh, Somerset. He owned Poulett House in Pound Street (now the Alexandra Hotel). The family stayed in Lyme Regis from October 1831 to 1833, when Miss Pinney was not yet 20. It was a lengthy residence in order to campaign for her brother William in the forthcoming elections, when he would run as a Reform candidate. Mary's brother, Joseph, was also a supporter of William Pinney. Lyme was one of the infamous 'pocket' or 'rotten' boroughs that would disappear with the 1832 Reform Act.

Anna Maria, with her lively mind, soon became acquainted with Mary Anning, accompanied her fossiling, and became a close confidante. Her journals show sides to Mary's character that she herself may have kept hidden, and are thus extremely valuable. However, they also reveal Miss Pinney's own character – moralistic, religious, gossipy and sentimental – and should be interpreted with some caution as the record of a privileged 18- or 19-year-old with her whole life ahead of her. When they first met, Mary was 32, by which time she had realised that many of life's chances were behind her.

Anna Maria Pinney was a visitor who recorded in her journals her personal impressions of Mary Anning. She first noted that Mary was 'a woman of low birth', but soon her comments became more perceptive as she fell under Mary's spell. On 25 October 1831 she wrote: 'She says she stands still, and the world flows by her in a stream, that she likes observing it and discovering the different characters which compose it.' Mary Anning was discontented with her own people, Anna Maria recorded. 'But in discovering these characters, she takes most violent likes and dislikes . . . Associating and being courted by those above her, she frankly owns that the society of her own rank is become distasteful to her, but yet she is very kind and good to all her own relations, and what money she gets by collecting fossils, goes to them or to anyone else who wants [needs] it.'[37] Mary seemed to be in between, neither one nor the other, and that is where she would remain. It must have grated deeply for Mary to watch these monied, educated gentlemen indulging themselves in her own field of expertise, with little or no thought of the expense. And she would have noted the very comfortable lives of their wives. Mary knew very well what she was missing.

*   *   *

Then there were the Philpot sisters, who were friends and colleagues in spite of their higher social standing. In about 1805, four sisters of a London solicitor moved to Lyme to search for fossils – and three were to remain in Lyme for the rest of their lives. The sisters were Mary (1777–1838), Margaret (?–1845) and Elizabeth (1780–1857), and all were serious fossil collectors.[38] They lived in Morley Cottage near the top of steep Silver Street on the right-hand side at number one, with a view over the town and sea, and the junction with Pound Street opposite.[39]

When the Philpots arrived in Lyme Regis, Mary was about 6, so they were becoming established at the time Mary was growing up and began to make her great finds. Although Elizabeth, the youngest of the sisters, was almost twenty years older than Mary, a close and affectionate relationship was to develop between her and 'the fossilist of this place'.[40] Mary and Elizabeth, an acknowledged expert in her own right, went out collecting together almost every day.[41]

The Philpots amassed a large collection of fossil fish, which they displayed in cabinets, and viewing the Philpot 'museum' was on the itinerary of almost every prominent geologist who visited the little harbour town, tucked so prettily in its combe. A local woman, Miss Selina Hallett (b. 1839), recalled in June 1920, when she was 81, that 'they [the Philpot sisters] were great fossilists and had a very large collection. Several cases with glass tops and shallow drawers all down the front stood in the dining-room and the back parlour and upstairs on the landing, all full of fossils with a little ticket on each of them.'[42]

Mary and Elizabeth well knew that there were certain places where one was most likely to find specific fossils or where shells might be washed up. Together they found innumerable examples of coiled ammonites, especially near Monmouth Beach, and walked on the exposed pavement of ammonites – some several feet wide – on the way to nearby Devon Head, passing the point where the Devon–Dorset county boundary runs down to the sea. In the other direction, to the east of the Cobb, was the old village of Lyme, where Mary grew up; beyond it and below fossil-rich Black Ven cliff they would have collected small, coiled ammonites and the bullet-shaped shells of belemnites. And this is where the fossils of the great reptiles were found.

Miss Guilelma Lister, one of the prominent Lister family, Quakers who from 1871 lived at High Cliffe[43] in Lyme, originally a holiday home for each of their families then a permanent home for her father Arthur Lister until he retired to his London home, mentioned the Philpots' 'famous Salve' in her recollections of Lyme.[44] Her father was a botanist and brother of the famous surgeon Joseph Lister, a pioneer in antiseptic surgery, who often visited Lyme and became 1st Baron Lister of Lyme Regis. Guilelma assisted her father in the field and in the laboratory. Although a fine naturalist–botanist herself, and a talented artist, she was dominated by her father.

Elizabeth Philpot was a serious collector, and in a letter of 1828 (to Charlotte Murchison) Mary presents a picture of them working together: 'You and Mr De La B [Beche] between you have given me green sand mania, both as Miss Philpot and myself have been beating bits of green sand to pieces to find shells, and after our return we sit down and turn over the leaves of Sowerby [James Sowerby, *Mineral Conchology of Great Britain*, 1812–29, in which he included a plate of one of the Philpot's ammonites in 1817] until our poor heads is a complete jumble of bivalves, univalves, etc.'[45] In a matter of weeks she wrote, 'I soon lost the green sand mania, and returned to my old passion for bones.'[46] They had been searching for fossils in the greensand, a sequence of sandstones, occurring above the Lias. From the Cretaceous Period, greensand has a sandy feel and contains the yellowish-greenish mineral glauconite.

*   *   *

Another important friendship was with Frances Augusta Bell, a young invalid, whom Mary met in 1824, when the latter was brought to Lyme for her health by her mother and aunt. Her letters to Mary, preserved in the girl's

THE PHILPOTS' SALVE

It was usual at the time and for many years afterwards for women to keep recipes for common illnesses in the most handy place, along with their prized recipes for preparing food. *Mrs Beeton's Book of Household Management*, 1860, was but one example, with seventy-five pages on treatments for childhood and adult afflictions. Another was the later *Things a Lady Would Like to Know* (written by a man, Henry Southgate), which contained, under the heading 'Recipes for Accidents or Slight Physical Derangements', recipes for treating ailments from gout to measles and chapped hands to toothache. For example: 'For Bleeding of a Wound: Take ripe puff-balls, break them warily, and save the powder; strew this on the wound and bind it on.' For a cut: 'Bind on melted cheese.'

The Philpots concocted their own soothing salve to be used on 'wounds', and no doubt Mary herself was the beneficiary of a constant supply. Selina Hallett wrote: 'they were very good, kind ladies to poor people. One thing they were noted for locally was their home-made salve, of which they always kept a store for anyone that chose to go and ask for it, and they had a great many applications for it, as it was so good for any sort of wounds.'*

There are two recipes for healing salves in *The Household Guide or Domestic Cyclopedia* (1894) that may suggest what comprised the Philpot salve. In one (for burns, frostbite, chapped hands) equal amounts of beeswax and sweet oil are melted together and turpentine is added. The other (for broken breasts, abscesses, fever sores) is composed of lard, resin, beeswax and tobacco. The range of information incorporated in these household books could range from how to raise canaries and a recipe for boot-blacking to how to eat oranges.

* J.M. Edmonds, 'The Fossil Collection of the Misses Philpot of Lyme Regis', *Proceedings of the Natural History and Archaeological Society*, 98 (1978), p. 47.

posthumous memoirs, refer to her interest in mineralogy over the previous three years, and how at Lyme she had met 'a congenial spirit' – Mary Anning – who was 'a perfect mistress of the science of mineralogy and fossils'. Mary's discoveries are recounted, and 'she is much noticed by the ladies of the place'. Frances Bell also recorded that Mary was modest and humble in a way that could not fail to please anyone who knew her.

The 25-year-old and the 15-year-old, both craving companionship, formed a friendship. Mary allowed the young invalid, whose medical complaint was a greatly swollen arm, indicative of a serious but unidentified disease, to accompany her in her daily walks on the foreshore.

The letters reveal a softer side to Mary's personality – motherly, emotional and religious. From Lyme in September 1824, Frances Bell wrote to a friend, Anne, that it was a useful place to collect shells, 'but an excellent one for fossils', and that she was busy forming a collection for herself as well as a duplicate one for her friend. After her stay at Lyme, Frances sent Anne her gift of the duplicate collection of fossils, and lists what was enclosed, giving us an idea of the diversity that could be collected in two or three months, albeit under the guidance of an expert who also generously added to the collections. It contained:

1. A part of the jaw of an icthyosaurus, a large fossil animal between the lizard and the crocodile
2. Vertebrae of the same animal
3. A piece of the rib of the same
4. Bezoar stone [coprolite], which admits of a beautiful polish

5. A piece of a dorsal fin. Animal not yet discovered entire
6. Scales of a fish
7. Ammonite, vulgarly called cornu ammonsis Species vulgaris. Ditto, mass ammonite
8. Ammonites stellaris
9. Belemite, a shell fish. Ditto mass
10. Gryphite. Species gryphea incurva
11. Terebratulae. Species anomia
12. Mya and cockle
13. Pentacrinite, a zoophite; the connecting link between the animal and vegetable world. One specimen impregnated with iron pyrites
14. Calcedony crystallized
15. Mammillated chalcedony
16. Amethystine, or polishing chalcedony
17. Selenite, or fluat of lime
18. Calcaeous spar
19. Quartz and iron pyrites
20. Iron pyrites and ball
21. Grape stone, or radiated iron pyrites
22. Mushrooms, impregnated with iron pyrites
23. Granite
24. Gneiss
25. Toadstone
26. Shell
27. Wood, shells and iron.

The Bells left Lyme in mid-October 1824, about four months after they had arrived. In November there was an exchange of letters with Mary. Although some of Mary Anning's letters to scholars survive, there are tantalisingly few personal remarks, so this exchange with Frances gives a sense of one aspect of her as a person. Miss Frances Bell had turned 15 on the very day she wrote from Highgate, 'near London', and in a note in her diary written on the same day – 23 November 1824 – recorded that she had only two friends and correspondents. One was unnamed and the other was Mary Anning, 'a poor, but intelligent and deserving girl'. The note continued: 'Which of these two will prove the most faithful, and abide by me the longest, time will shew. I am resolved to do my duty to both.' Then she wrote her belated letter to her kind older friend, Mary Anning:

My very dear Mary,
A whole month having elapsed since I left you, and dear Lyme, without your hearing from me, I fear you will begin to think me unkind in not fulfilling my promise of writing, or suppose that illness has prevented me. My fossils escaped a smash, and were very much admired; particularly the metalized ammonite you gave me, and the mineral from which you take your title; indeed I was almost scolded for not bringing home all I found of it, (a pretty undertaking! you will join me in exclaiming,) for they were not more than half inclined to think it was gold . . . My shells also, which I have with me, traveled delightfully; and those you kindly gave me are very much admired. I assure you, that where we are, there you will not be long unknown . . . We bore our journey tolerably and arrived in smoky London at nine at night. O how different from the pure air you breathe!

She also gave Mary a brief medical report: 'we went the next day to Dr——'s, and from his prescriptions have received great benefit . . . My arm is smaller, and much stronger: the bath is very useful.' Knowing what would be of most interest to her friend, Frances continued,

I have twice visited the British Museum; and intend devoting a day to each room, in order to examine its beauties more minutely, and to improve myself in the study of minerals, shells and fossils, as much as I can, by so good an opportunity of seeing the true specimens. O! Mary, you never saw nor can conceive, any thing so beautiful as are the minerals. I never go there without wishing for you to participate [in] my pleasure.

They have also a good collection of shells and fossils, in fishes and in leaves, but not in the grander specimens; I mean animals and ammonites. Although the head is larger, and the eye more conspicuous, I cannot say I admire the ichthyosaurus you first sent there half so much as the one you now have. It is styled immanis; do explain this. The marbles I admire, as I should any other fine specimens, from their antiquity, and the classical remembrances to which they naturally give rise . . .

She concluded, 'Pray write soon, and tell me if I have not tired you; if not, I will give you an account in my next of what else I shall have seen. "Heaven's blessing be with you!" were your emphatic parting word: it has, dear Mary; and that it may always continue with you, is the constant prayer of Your very affectionate friend, F.A. Bell.'

Six days later Mary received the letter. She replied quickly, on 29 November 1824. From Cockmoile Square, where she lived in 1824, it was a short walk for Mary up narrow Combe Street to the post office. Here she posted back her letter in the timber letterbox set into the outside wall, where it remains to this day. It was a walk she had made many times, for the Congregationalist Chapel was almost next door, further up the street.

My dearest Fanny,

Many thanks for your kind, interesting letter, and I have to beg your pardon for doubting your friendship; not hearing from you for six weeks instead of two, I thought if illness had been the cause of your silence, your dear good aunt would have sent me one line, just to tell me: *the world has used me so unkindly, I fear It has made me suspicious of all mankind*, I hope you will pardon me, although I do not deserve it. Dear girl, how rejoiced I am that your arm is stronger and smaller! You did not say how you or dear mamma's health and spirits were; mind you do it in your next.

How I envy you your daily visits to the museum! indeed I shall be greatly obliged your sensible account of its contents; for the little information I get from the professors is one-half unintelligible. I cannot explain the term they have given to this skeleton, unless it means approaching the lizard tribe. Very little doing in the fossil world; excepting, I have found a tail for baby, and a beautiful paddle, and a few other small specimens; nothing grand or new.[47]

Portraits of the adult Mary reflect her forthright, stern personality, revealed in her neat, no-nonsense appearance. Two very similar images/likenesses of a full-frontal view survive. In one, oil on board, she is pointing to an ammonite. She seems somewhat insipid, until we look again, bearing in mind all that she accomplished. One notes the clear penetrating eyes, the high intelligent forehead, the strong nose and chin and the self-contained demeanour. She was of medium height, thin, with dark hair. To a child her eyes were large and kindly.

At first glance her face looks bland, not beautiful, not ugly, but 'nice' – an acceptable depiction of a respectable woman in an unusual role. She was described as being 'masculine' in appearance. The all-enveloping cloak concealed a body described as slight, but she must also have been physically strong and sturdy. Once, when Anna Maria Pinney was out fossiling with her, they were caught by the tide on a very narrow space between the sea and the rock, and had to dash across between two waves. She carried Anna Maria to safety: 'before I knew what she meant to do, she caught me with one arm around the waist, and carried me for some distance, with the same ease as you would a baby.'[48]

The age of photography was on the horizon but had not yet arrived in Dorset. There is no photograph of Mary Anning, although Selina Hallett in her recollections thought her mother had had a photograph, but did not know what had happened to it. A portrait of her in oils was probably painted by W. Gray in February 1842 for an exhibition at the Royal Academy (but rejected, which perhaps says something about the quality of the painting). It was then offered for sale to the National Portrait Gallery, then the British Museum (Natural History) 'during Sir A. Smith Woodward's Keepership', then to historian and author Cyril Wanklyn of Lyme. All refused it. Finally, it went to her brother Joseph. In 1935 it found its proper home beside some of the important fossils she discovered when it was presented by Mary's great-great niece Annette Anning (1876–1938) to the Department of Geology, British Museum. It is now in the Earth Sciences Library, at London's Natural History Museum. A large ammonite is there, as is her terrier, Tray, dutifully curled up resting, looking somewhat like an ammonite.

A very similar portrait in pastels was painted by B.J.M. Donne (1831–1928). This likeness relies heavily on the one by Gray. Although painted in pastel in 1850, when the artist was only about 18, and around three years after Mary Anning's death, Donne had attended George Roberts's school in Lyme and must have known her. This portrait was commissioned by the group who donated the stained-glass window to her memory in St Michael's Church in Lyme. It is now at the Geological Society.

There is mention of another portrait by 'Miss Wyse', sister of Sir Thomas Wyse (MP for Waterford, who married Napoleon's niece, Letitia), who painted the portrait for a museum in Ireland. 'It is perfectly unlike' was the handwritten note in Roberts's book. This portrait has never been traced. Also, in about 1838, a private lady painted a portrait in watercolour for a Miss Penning. 'It is very like' was the comment, but it too remains undiscovered.[49]

An illustration that has a ring of truth about it is a watercolour sketch of Mary in her more mature years by her old friend Henry De La Beche, who skilfully sketched and caricatured most people in his circle. It depicts her at work on the beach with her geological hammer, layers of clothing, check shirt over petticoats, sturdy boots, and a battered high hat, possibly a forerunner of a hard hat, all of which make her look heavy and somewhat mannish.

In the likeness her hammer is in her right hand and her rush collecting bag is slung across her back. Unfortunately, it is a rear view.[50] An anonymous contributor to *Chambers's Journal* in 1857 confirmed again that she was of 'rather masculine appearance'.[51]

Her hats in these images were either the poke bonnet or the tall masculine 'hard hat'. Plaid, no doubt woollen, seems to feature somewhere in her clothing in each portrait. We are left with an image of a sturdy, unglamorous, determined woman with a strong independent streak and an enquiring mind.

While much was 'wrong' with her life, at least one thing was right: Lyme Regis. There, more monsters were suspended in time waiting for her keen eye to winkle out the hidden hints and clues as to their resting places. The carpenter's daughter continued to reaffirm her own impact on the new geology.

3

Denizens of Primeval Waters

a monster resembling nothing that has ever been seen or heard of upon earth, excepting for the dragons of romance and heraldry.

(Cuvier on the pterodactyl *Dimorphodon macronyx*[1])

On 10 December 1823 Mary made her second major discovery in the shale at the foot of Black Ven cliff: the first almost complete articulated *Plesiosaurus* ('near to reptiles'). After finding the skull, she worked through the night in freezing conditions to release the vertebrae, ribs, pelvic bones and the fine bones of its four paddles. It was 9 feet long, but the head was only 4–5 inches in length.

That Christmas of 1823 she must have quietly given thanks, because the appearance of this important new type of marine reptile indicated a comfortable winter ahead. With this discovery, Mary became well known enough to have a London agent who would sell her fossils. George Brettingham Sowerby was the son of the botanical artist, James Sowerby. This dealer in natural curiosities became Mary's agent the following year, taking 20 per cent in commission. Mrs Murchison also promoted Mary's skills and finds, and gave her a list of specimens they themselves required.

Back in the winter of 1820/1 Mary had found her first plesiosaur in the soft marl by the cliff, but it had presented problems. There was no skull, and furthermore the brittle skeleton was in pieces by the time she got it back to her workroom, so, as she was reassembling it, it was not absolutely clear how one part related to another. However, the plesiosaur was quite unlike an ichthyosaur or indeed any other creature she had ever seen. That first plesiosaur must have had a very small head, she surmised. Although the skull was missing, the slender snake-like neck as long as its body could only have supported a compact skull (in contrast to the ichthyosaur with a relatively large head, no neck, and somewhat resembling a dolphin). It was memorably described by Buckland at the time as 'a serpent threaded through a turtle'. She had sold the first incomplete specimen to Colonel Birch, probably for about £100 or less.

By the time of this second – and much more important – *Plesiosaurus* find late in 1823, Mary Anning was already famous on several fronts: for her childhood discovery of the *Ichthyosaurus*, for the dramatic sale in London by Colonel Birch of his collection of fossils purchased from her, and for the 1821 discovery of a fine small ichthyosaur, *Ichthyosaurus communis*. Two months later had come another ichthyosaur. And in 1822 she had found yet another. She was well known and respected among the gentlemen scholars to whom she was supplying fossils, so word of this latest discovery was received with great excitement in London. A newspaper report declared, 'the well-known fossilist . . . found . . . immediately below the celebrated Black Ven Cliff, some remains which were removed on that night and the succeeding morning, to undergo an examination, the result of which is, that this specimen appears to differ widely from any which have been before discovered at Lyme . . . while it approaches nearly to the structure of the Turtle . . .'.[2]

Buckland adopted Cuvier's description: 'To the head of a lizard, the plesiosaurus united the teeth of a Crocodile, a neck of enormous length resembling the body of a Serpent, a trunk and a tail having the proportions of an ordinary quadruped, the ribs of a Chamelion, and the paddles of a Whale.'[3] The four powerful diamond-shaped paddles moved in a figure-of-eight pattern, giving the creature a deft and strong locomotion that was a cross between flying and rowing. (Descriptions of the ever elusive Loch Ness monster seem to have been inspired by that of a plesiosaur.)

The find of the *Plesiosaurus giganteus*, a creature that was about 200 million years old, drew closer into Mary's circle yet another prominent pioneer from the fields of geology and palaeontology: Revd William Daniel Conybeare. The leading authority on both ichthyosaurs and plesiosaurs had already named it *Plesiosaurus* back in 1821. Conybeare, twelve years older than Mary, was already a good friend of De La Beche. They had first met at the Assembly Rooms in Lyme,[4] were at Oxford at about the same time, and had worked together on both the genus *Ichthyosaurus* and the genus *Plesiosaurus*. Before the almost complete 1823/4 specimen was found by Mary, they had already identified the genus from a selection of incomplete finds, and had written about this new type of marine reptile that lived at the same time as *Ichthyosaurus*. This jumble of small fragments came from different collections, some in the Bristol area, only about 80 miles from Lyme; it was quite possible that some of them had been found initially by Mary. And Birch, who had bought the first incomplete plesiosaur found by Mary, allowed Conybeare to examine it. Even Conybeare – who had described it, but with some guesswork – at that point still had doubts as to whether this really was a new kind of 'sea lizard' that had existed at the same time as the *Ichthyosaurus*. These two genera, along with marine crocodiles, were the dominant animals in the Jurassic sea.

Now, in 1823, the eminent gentlemen had an almost complete specimen of *Plesiosaurus*, with barrel-shaped body, basketwork frame of abdominal ribs and small head. Once again, it was Mary Anning, not the scholars, who had made the breakthrough find. And Thomas Hawkins in his *Memoirs* praised her to the skies: 'Miss Mary Anning to the due expression of whose public and private worth all language is insufficient . . .'.[5] But it was Conybeare's name – since he had already named and described it – that would become part of the scientific identification, as was the custom. There was justification in him doing this, as he had done much work on the genus before Mary's find. What is unacceptable is the way in which she was cut out from any official recognition.

Mary's complete *Plesiosaurus* was first recorded by Conybeare at Bristol, in a letter to De La Beche who was in Jamaica at the time. On 4 March 1824, he wrote to tell him the 'important news' that he had just received from Buckland, who had dropped in on his way to Lyme

Regis. In the excitement the deeply devout Conybeare – perhaps symbolically – was torn away from the sermon he was writing, leaving one of his sisters-in-law to finish it for him. Buckland described the competent and detailed pen and ink drawing of the skeleton that Mary had made. Conybeare continued in his letter to De La Beche, 'I begged him [Buckland] to send me immediate intelligence thence, and three days afterwards I received a very fair drawing by Miss Annin [*sic*] of the most magnificent specimen . . .'.[6] He took the drawing of 'the infant nursling of my own' – immediately laying intellectual claim to the specimen – to that evening's meeting of the Bristol Institution. There it caused so much excitement that some people ran off to the newspaper office and he was obliged to follow them there to prevent misinformation being circulated.

Conybeare had already arranged to set off on a journey the following day, and now Buckland asked him to include London in his travels as well. There were good reasons for this: the next Friday was the meeting of the Geological Society at which Buckland was to be elected president, it would be thrilling if Mary's plesiosaur was there for the eminent Fellows to examine, and Buckland wanted his trusty friend Conybeare to be present when it arrived in London.

The titled purchaser of the plesiosaur was Richard Temple-Nugent-Bridges-Chandos Grenville, the 1st Duke of Buckingham and Chandos (1776–1839). The main accomplishment of the Duke's life was securing the dukedom for his family in the reign of George IV. Immensely fat, greedy and ambitious, he was involved in blackmail, and altogether a deeply and incorrigibly debt-ridden MP. He could not have been less popular. By 1827 he was sent abroad to save money, but his 'economies' included having a yacht built with extra wide gangways, and the result was more debt than ever. There was little to commend him, except his collecting.

The Duke had charged Buckland with the care of the specimen, in order to complete a scientific investigation of it and confirm that the creature was what it was supposed to be. For this purpose it was shipped to the Geological Society in London, where the Fellows were soon due to meet. But it was now that the real tussle among the Fellows began. Buckland had asked Conybeare to be present when Mary's plesiosaur arrived in London to keep it out of the grasping hands of the eminent Sir Everard Home, who was prominent in both the Royal Society and the Geological Society. Sir Everard Home (1756–1832) was the son of a surgeon, who had become the first president of the Royal College of Surgeons. Home junior made up for being neither clever nor industrious by being vain and ambitious. A friend of the Prince Regent, he controlled the Royal Society with a tight rein. Home was not trusted by many in the elite geological circle – and based on past performance, he was the reason Cuvier was initially dubious about some of the English finds.

Sir Everard's sister, Anne, a poet, had married the great surgeon John Hunter, known as 'the father of modern surgery'. This familial link gave Home considerable influence. Hunter made Home executor of his will, thus, on the former's death, Home was left in control of the large number of Hunter's valuable and unpublished papers. These he removed from the Royal College of Surgeons, supposedly to catalogue them. He then began publishing Hunter's papers under his own name. To protect himself, he copied them out, then reputedly burned most of Hunter's original research in about 1823, having kept it for thirty years. However, when William Clift, Curator of the Hunterian Museum, Glasgow, visited Home, he noted to his horror that some of the priceless writings were being used as toilet paper!

Home himself had been studying the creature that became known as *Plesiosaurus*. His theories were based on fossils in William Bullock's Museum of Natural Curiosities, in the Bristol Philosophical Institute, and in the collection of a Mr Johnson of Bristol, who had obtained at least some of his collection at Lyme. And, according to Mantell – referring to Bullock's collection – 'All the most valuable fossils had been obtained by the indefatigable labours of Miss Mary Anning.'[7]

When Conybeare reached London in February 1824, the specimen had not yet arrived because it was delayed in the Channel; the massive shipment did not appear on the doorstep for a further ten days, and so missed the important meeting at the Geological Society. When the gigantic fossil finally arrived at Somerset House, a whole day and much energy were lost trying to get it upstairs for unpacking. The packing box was 10 feet long and 6 feet wide, and simply would not budge further than the entrance passage. Here, the unpacking finally began, and here it would be examined by the Fellows in flickering candlelight.

It was hard work trying to keep the eminent Sir Everard out of things. Typically, he made strenuous efforts to be the first to announce a new genus or species. With Mary's creature, he rapidly proclaimed its discovery, even naming it *Proteosaurus*, asserting it to be a link between the *Proteus* (tailed amphibians with gills, short legs and an eel-like body) and lizards. The name was completely inappropriate once the entire skeleton was understood and the name *Plesiosaurus* was accepted. Home, however, determinedly stuck forever with the name he had given it. (Home had caused even worse confusion when Mary and Joseph found the ichthyosaur back in 1811–12, insisting that it was a crocodile, then a fish, or a new genus covering both, and finally, perhaps, an amphibian. He had not even guessed at a reptile.) Everyone knew that Mary Anning had found the plesiosaur, but she was, as always and without a second thought, sidelined.

Although the fossil itself was not available for the Geological Society meeting on 20 February 1824, Conybeare was immensely pleased with Mary's drawing of the ichthyosaur because it confirmed emphatically his own theories. Chemist, physicist and inventor William Hyde Wollaston (1766–1828) (founder of the medal for mineralogical research) was so keen, he reduced the drawing using the camera lucida, which he had invented. De La Beche had already prepared a set of detailed drawings of the plesiosaur, 'a new fossil animal', a link between the ichthyosaur and a crocodile, while Conybeare would describe it in *Transactions of the Geological Society of London*.[8]

Conybeare entitled his talk to the Geological Society 'Notice on the Discovery of a Perfect Skeleton of the Plesiosaurus' – Mary's plesiosaur. The Fellows marvelled at news of the creature's numerous neck vertebrae, how it swam near the surface, neck arched like a swan (later proven not to be true), and how its paddles had ten bones in each long digit. It was turning out to be a successful and most enjoyable evening. Ungallantly and disappointingly, Conybeare said that he had first misinterpreted the giant fossil because of the way the 'proprietor' had assembled the pieces. The unnamed 'proprietor' was Mary Anning. There was no wish to include her, or credit her with the discovery, but she was handy when there was blame to be apportioned. It was clear that he intended to blot her out of the story. At the end of his lecture he announced its name for the first time: *Plesiosaurus giganteus*.

However, on this extraordinary February evening – did any of those attending ever forget it? – the thrill of hearing Conybeare's lecture was almost immediately overtaken by William Buckland's following talk. A powerful and always entertaining speaker, Buckland – marking the first meeting at which he was President of the Society – announced *his* discovery of a *Megalosaurus* at Stonesfield quarry, named by himself in consultation with Conybeare. He revealed that this carnivorous reptile, popularly called the Stonesfield giant, had reputedly been 60–70 feet long (later corrected to 30 feet). Buckland tried to keep it simple by saying it was much like modern lizards, just bigger. In fact, it was the largest terrestrial fossil from the Mesozoic Period found to date. No one realised it at the time, but this was the first dinosaur fossil discovered. It was, indeed, an evening to remember in the annals of the Society.

Mantell was in the audience and exceedingly interested. He had his own collection of *Megalosaurus* fossils, even larger specimens, some of which he had brought with him. Buckland soon visited Mantell to see his excellent collection. There was a suggestion that Buckland intended to incorporate Mantell's findings into his own published account, but he was warned against doing so by the Society.[9]

Stonesfield, near Oxford, was Buckland's most favoured quarry, but any rocky outcrop, cave or quarry was like a magnet to him – he could almost sniff them out. Buckland's olfactory organs did seem to be extraordinarily sensitive. Once, when he and a friend were riding towards London, they got lost in the dark. Buckland dismounted, scooped up a handful of earth, smelled it and correctly proclaimed: 'Uxbridge'.[10] His faithful old black mare, carrying him with his ever present blue collecting bag and a load of other bags of fossils and hammers, was so attuned to her master's whims that if she saw a quarry she would stop, and remain there without being held; and if a stranger was mounted on her, he had to dismount and pretend to examine the rocks, before the horse would move off again.[11] Likewise, the mare had observed that William and his wife, Mary, were very fond of flowers; on their Sunday drives when they came to a nursery, the attentive quadruped stopped until an appreciation of the floral abundance had taken place.[12] Anyone riding in Buckland's specially strengthened carriage might gradually be edged out by the rocks that one of the leading geologists of the day collected along the way wherever he went, and find themselves obliged to walk alongside.[13]

Soon, a controversy over Mary's plesiosaur added to the general exhilaration, when Cuvier, on being shown a drawing of the new creature, declared that he suspected it was a fraud. With a frisson of excitement, a meeting was convened at the Geological Society to discuss the question. Mary Anning was not there. Eventually, Cuvier acknowledged that he had been wrong. Even a leading vertebrate palaeontolgist was not infallible. This episode only added to Mary's fame, with the result that the

Anning name was henceforth treated with respect in the scientific community.

William Daniel Conybeare had first become interested in geology at Christ Church, Oxford, with William Buckland, and there he had met De La Beche in 1818. He took Holy Orders, and his second posting in 1836 could not have been a coincidence – he became Vicar at Axminster, where Buckland grew up, which was only 6 miles inland from the fossils at Lyme and Charmouth; he remained there for the next nine years. He wrote several important papers, as well as the greatly influential (with William Phillips) *Outlines of the Geology of England and Wales*, the first widely used work on geology in England; geologist Rodney Impey Murchison later called it his 'scientific bible'. Buckland was one of the most famous of the early geologists, but Conybeare was intellectually the most impressive according to Nicholaas A. Rupke, author of *The Great Chain of History: William Buckland and the English School of Geology (1814–1849)* (1983).

Six months after Mary's discovery of the plesiosaur, Edinburgh mineralogist Thomas Allan (1777–1835) visited Lyme, and on 25 June 1824 recorded:

Mary Anning the Geologist of this place is a very interesting person, and the scientific are entirely indebted to her for the preservation of some of the finest remains of a former world that are known in Europe. For the large specimen of Plesiosaurus now at the Geological Society of London, and for which the Duke of Buckingham paid her £100 [prices quoted ranged from £100 to £150 and £200], we are entirely indebted to her industry – and it was our good luck to find her engaged on another specimen of the same kind – equal in size but a different animal – which she had discovered only last week . . . Mary Anning's knowledge of the subject is quite surprising – she is perfectly acquainted with the anatomy of her subjects, and her account of her disputes with Buckland, whose anatomical science she holds in great contempt, was quite amusing.

Thomas Allan continued, 'She walked out with us on the beach, and showed us where she looked for and found her best specimens.' As they strolled, eyes down, she found a very fine dorsal fin of a *Hybodus* (shark), for which he gave her half-a-guinea. 'She says she is indebted to her father for all the knowledge she has . . .'[14]

In late 1824 disaster struck Lyme when the Great Storm mercilessly pounded the town, along with many other communities along the south coast of England. Clearly upset, Mary described the event in a letter to the young invalid, Miss Bell.

Oh! My dear Fanny, you cannot conceive what a scene of horror we have gone through at Lyme, in the late gale: a great part of the Cobb is demolished, every vessel and boat driven out of the harbour, and the greatest part destroyed; two of the revenue men drowned, all the back part of Mrs England's houses and yards washed down, and with the greater part of the hotel [England's hotel at the Cobb], and there is not one stone left of the next house: indeed, it is quite a miracle that the inhabitants saved their lives. Every bit of the walk, from the rooms to the Cobb, is gone; and all the back parts of the houses, from the fish-market to the gun-cliff, next the baths. My brother lost, with others, a great part of his property.[15]

This may imply that her brother Joseph was no longer living in the family home, or perhaps it was the point at which he gave up his interest in fossils. In the mid-1820s, he became an upholsterer and on 28 November 1829 married Amelia Reader (c. 1797–1858). Five of their children were William (1833–8), Charles (1834–adulthood), Joseph (1836–8), Prudence (1838–adulthood), at least one other child who died in infancy, and Albert (1842), through whom the line carried on when he married Annie Jerrard in 1866. Joseph continued to care for his mother and sister after he married, for records show that he paid two rents. He became something of a stalwart of the community, he was elected to the Vestry of St Michael's and in 1835 became an Overseer of the Poor.[16] The first fossil collection in Lyme may have been at the Working Men's Club where Joseph Anning was a member.[17]

In her letter concerning the storm Mary continued: 'All the coal cellars and coals being gone, and the Cobb so shattered that no vessel will be safe there, we shall all be obliged to sit without fires this winter: a cold prospect you will allow.' The raging water had dragged the sailing ship *Unity* out of the Cobb, with the terrified crew still clinging to the rigging. This ship had sailed regularly with goods between Lyme and London, where Captain Smith's wharf was set on the south bank of the Thames below old London Bridge in what is known as the Pool of London. But the *Unity* would sail again. She was swept ashore not far away – under Black Ven. The crew escaped with their lives, and the boat was repaired to sail for many more years. Mary later sent a 4½-foot long *Ichthyosaurus* specimen[18] to Sedgwick at Cambridge on the *Unity*, by then under the command of Master Pearce.

Mary wrote to Sedgwick on 9 September 1835 to inform him that it was on the way.[19]

So hard hit was Cockmoile Square that Mr Bennett, the Annings' long-standing neighbour, and others paid to have the sea wall built up once again, as they had done before, not realising this was the responsibility of the town and the Lord of the Manor. A complicated court case resulted involving Henry Hoste Henley, the Lord of the Manor, who owned much land in and around Lyme and had not kept in good repair those sea defences for which he was responsible. During these debates it came to light that a ship had once been built at that spot at the back of Cockmoile Square where part of the wall had had to be removed in order to launch it. Over the years there had been shipbuilding at several sites in Lyme, where merchant craft 'as handsome vessels as ever swam' were constructed.[20]

Bennett's other neighbour was a Mr Blackmore, who ran a school there. His part of the wall was totally destroyed and the rest of his garden fell into the sea. Apparently, a toilet sited there annoyingly had deposited waste on the steps to the foreshore but, happily, during the storm was itself deposited in the ocean.[21]

In 1825 Mary watched the Cobb – the most prominent feature of Lyme Regis then and now – being rebuilt. This time it was constructed entirely of limestone from the quarries on the Isle of Portland south of Weymouth, only a few miles away. And at this point the characteristic steps known as 'Granny's teeth' were built. She remained the one with an eagle eye and an instinct for spotting where a fossilised monster might be buried just out of sight. When Mary was in her mid-twenties, a visitor had commented: 'Such is her intelligence that most visitors to Lyme request to accompany her in her walks of science.'[22]

Also that year Mary Anning was visited by the young geologist Roderick Impey Murchison (later Sir) (1792–1871). He had retired from the army after Waterloo and settled down to raise a family. His wife, Charlotte (1788–1869), a future heiress, encouraged him to take up a new pastime and suggested geology, to extricate him from the unchallenging role of a fox-hunting country gentleman. It worked, and at once he threw himself into his new interest with his usual gusto.

Murchison was destined to become the most famous pioneer of Silurian geology (an early Palaeozoic System of rocks (443–416 million years old) lying immediately below the Devonian System). With geologist Adam Sedgwick he established the Devonian System (416–360 million years old). Years later he described his first idyllic fossiling expedition with his beloved Charlotte, another remarkable woman who has her own entry in the *Dictionary of National Biography*: 'upon my beloved wife's influence over me for whatever good I may have done in the walks of science . . . Immediately I had acquired this taste for Geology my wife and self resolved to explore for ourselves (1825) the whole southern coast of England, from the Isle of Wight inclusive, where all our home phenomena were repeated, to Land's End.' He tells how they travelled: 'For this purpose I had a nice little pair of horses, a light carriage, and with saddles strapped behind to use the nags for riding when at any centre of attraction. At some places we examined the cliffs in boats, she never failing to make good sketches.' By the time they arrived at Lyme Regis, Charlotte was tired, so he left her there to 'become a good practical fossilist, by working with the celebrated Mary Anning of that place, and trudging with her (pattens on their feet [wooden platforms to keep above the mud]) along the shore; and thus my first collection was much enriched'.[23] He and Charlotte were to become true friends to Mary.

Another eminent geologist who had a clever wife was Charles Lyell. In 1832 he married Mary Horner, the daughter of a geologist, in Germany. They went on a geological honeymoon before settling in London near the British Museum. Fluent in several languages, she was a great help to him in translations and documentation. However, he excluded her from discussions about science, behaviour that apparently annoyed Charles Darwin.

*   *   *

Geology and geography framed Mary's life. Although interest was growing in geology and fossil collecting, the latter was not a common activity for a man to undertake and certainly not for a woman. For a female to be associated with the profession of geology and employed in selling fossils was considered deeply shocking by many – ladies simply did not involve themselves in commercial activities. In 1832 Maria Hack (1777–1844), an educational writer, noted in *Geological Sketches and Glimpses of the Ancient Earth*: 'It is certainly uncommon to hear of a lady engaging in such a fatiguing, hazardous pursuit; and I think few would be found willing to undertake a personal examination of the cliffs, especially in the depths of winter.'[24]

However, her fame continued to spread. In 1824 Maria Pinney had noted, 'all the professors and other clever men' in geology 'acknowledge that she understands more of the science than anyone else in the kingdom'.[25] And, in October 1827, an American geologist arrived at Lyme. George William Featherstonhaugh (1780–1866) was collecting fossils for the recently opened New York Lyceum of Natural History. Although the specimens that he collected have since disappeared, we are left with another memorable description of Mary. To him she was 'a very clever, funny Creature'.[26]

At a time when a woman did not walk out with a man to whom she was not related, Mary Anning was visited frequently by many great scholars, all of them men. On their part, they were delighted while fossiling to accompany someone who had so much knowledge which was imparted so generously, and which they could then use to enhance their own careers and income. Mary was not overawed by these men, but she knew her class and her place. She was well aware of the fact that she had to be respectful, to a degree, to these learned gentlemen who were her betters socially, even those she counted among her friends. For Mary, it was a matter of economic survival. This is why the comment by Thomas Allan in his 1824 journal that Mary Anning held Buckland's anatomical science in great contempt rings true. She was not a saint, and the frustration had to erupt at times. Her friendship with her colleague Buckland was still sincere, and her enquiries about the health of his family, particularly the children, genuine.

In 1826 Mary and her mother moved away from Cockmoile Square, but not too far – just uphill, away from the sea, to the more prestigious 'Top of the Town'. Almost two years had passed since the Great Storm, and no doubt they had had more than enough of living so perilously close to the terrifying rages of the ocean. And they must have been relieved to be on safe ground when there was another serious storm later that same year. Their new abode was at the upper end of the main street, steep Broad Street, at 28 Haddons. This became home, workshop and shop. The move was reported in the *Sherborne Mercury*, 22 July 1826: 'Miss Mary Anning's Fossil Depot is removed to Broad Street where it will always prove a source of attraction.' The article added that there was a 'splendid animal' on display, an 'Icthyosaurus Tenuirostris'.

Little is known of Mary's mother except that she, too, was an 'original'. Sometimes the two were confused in written accounts, a point to bear in mind. And occasionally Mary had to apologise for her. In January 1830 Mary Anning wrote to J.S. Miller of the Bristol Institution, regarding fossils of interest and referring to a sharp-tongued error her mother had made: 'I am extremely sorry that I was not at home when yr Friend called, I beg ten thousand pardons for my mother's mistake. It was another Gentleman of the same name of whom I had complained . . .'.[27]

In their new abode they were visited by Nellie Waring (also known as Sister Emma). She wrote her memoirs at the end of the nineteenth century, and recalled visiting Mary and mother Molly in their shop: 'there lived this very timid, very unpretending, very patient and very celebrated woman, the discoverer of the Ichthyosaurus and of other fossil remains which were living animals before the Deluge.' It was a favourite place for children to visit. They were fascinated by the mysterious and beautiful fossils, and were always treated kindly by Mary. Nellie continued:

We, as children, had large dealings with Miss Anning, our pocket-money was freely spent on the little Ammonites which she washed and burnished til they shone like metal . . . She would see to us with the sweetest temper, bearing with all our little fancies, and never finding us too troublesome as we turned over her trays of curiosities, and concluded by spending a few pence only, and this we might do as often as we liked without offence.

A further description of Mary and the shop states: 'She must have been poor enough, for her little shop was scantily furnished, and her own dress always of the very plainest, she was very thin and had a high forehead, and large eyes which seemed to me to have a kindly consideration for her little customers.'[28] A Lyme woman pictured the shop in her memories of Lyme Regis in 1840: 'About half way down Broad Street, on the west side, stood a plain, unpretending little shop with a small white oblong board over the door, whose primitive and crude lettering informed the public that it was richly stored with precious specimens of the saurians [lizards], pterodactyls and other fossilised remains of the denizens of primeval waters, and it was a frequent resort of many of the most eminent scientists of the time.'[29]

A less kind description of the premises was given by the clever but unlucky Gideon Mantell, who may, in part, have been reflecting his own frustration. When he visited Mary in 1832, he 'found her in a dirty little shop, with hundreds of specimens piled around and in the greatest disorder'.[30] 'Dirty' because of the unavoidable grey dust created when chipping away at fossils as well as the cutting and polishing. And Mary did not have servants or even assistants to clear up after her, in what was, inevitably, a messy and dusty business.

Mary's shop is placed in context when compared with a wonderful description of William Buckland's rented rooms at Lyme. Could there have been any greater disorder than that which surrounded Buckland, both in his Oxford apartments and in his rooms at Lyme, described by his daughter, where

> his breakfast table at his lodgings there, loaded with beef-steaks and belemnites, tea and terebratula [brachiopods], muffins and madrepores [corals] . . . every table and chair as well as the floor occupied with fossils whole and fragmentary, large and small, with rocks, earth, clays and heaps of books and papers, his breakfast hour being the only time that the collectors could be sure of finding him at home, to bring their contributions and receive their pay.[31]

And Mantell's own abode was not much better. When the family lived in Brighton, to make money he turned their house into a museum bursting with an ever rising tide of skeletal exhibits and visitors. The family were literally forced out of their own home by thousands of bones.

*   *   *

The next extraordinary discovery came in 1828. Mary noted part of an invertebrate that had something very special about it. It was the 'anterior sheath and ink bag of a *Belemnosepia*'; in the bag was what turned out to be the ink of a gastropod. The ink of this fossil cuttlefish (*Sepia*) – the squid-like cephalopod, *Belemnosepia* – still

contained dried-up fossil ink (similar to the defensive ink
of an octopus or squid). After being macerated and
reconstituted, surprisingly the ink was usable. Mary gave
the sepia ink to Elizabeth Philpot, who, charmingly, used
it to draw Lyme fossils. Elizabeth Philpot wrote to Mrs
Buckland about it in a letter of 9 December 1833, and
enclosed a drawing of an unidentified fossil using ink
'prepared from the ink of the fossil Sepia', adding that
Mary Anning 'considers this jaw the most perfect she has
ever met with'.[32] The ink was also used to draw fossil
pictures sold in Lyme, and interest in them helped to
increase tourism. Buckland himself noted that 'a celebrated painter' said that the fossil ink was 'of excellent
quality',[33] while John Murray recounted that Mary told
him that Sir Francis Chantrey had once given her a drawing in sepia ink of an ichthyosaur.[34] Buckland announced
this latest find in an article entitled 'Fossil Sepia', in the
London and Edinburgh Philosophical Magazine of
1829.[35] Mary's latest discovery brought even more
publicity.

Years earlier Buckland had noted specimens in the
collections of the Philpots and Annings at Lyme with
what appeared to be a sac like a gall bladder containing a
hardened black substance. When powdered it was like a
painter's sepia. The 'ball or sac was enclosed in a bril-
liant nacreous sheath forming the lining of a thin shelly
substance which was prolonged into a hollow cone'.[36]
This and other features seemed to link it, in this case, to a
belemnite.

Mary wrote to Buckland on 21 December 1830: 'Do you
recollect the live creatures you had put in spirit (if so my
discription [*sic*] may be acceptable) I got two more, one
alive, and it, whenever it was touched ejected a purple

fluid (this one Mr De La Beach [*sic*; this is how it was pronounced] coveted, and had taken to some Naturalist to be described).'37

To visit her friends the Philpots, Mary turned left from her front door, walked uphill a few steps to where Pound Street and Silver Street meet, and turned right into steep Silver Street. She went to the Philpots to dissect one such specimen.

it first had a shell this shape [sketch] very like a smooth pecten only more concave, also a sac or ink bag exactly resembling the one you borrowed of Miss Philpot this shape [sketch] it had also a second small bag (like the gisard of a fowel [*sic*]) containing a number of horny triangular little pieces (I should think to assist its digestion) it had four horns in the two uppermost are its eyes; round the shell, I should have observed, a lip of fleshy substance which completely covers and gards it is [in] case of necessity.38

There were other references to prehistoric animals with ink sacs. Anna Maria Pinney recorded in her journal: 'Friday 7th [December 1832]. I went out fossiling with Miss Spekes &c &c. M. found a perfect Belemnite with its ink-bag – the only one previously found belongs to Miss Philpot.'39

All this clearly occurred before the first visit to England, in 1834, of the famous Swiss palaeontologist and naturalist Jean Louis Rodolphe Agassiz (1807–73), a protégé of Cuvier, under whom he had studied comparative anatomy. Through Mary's knowledge of fossil fish she made the acquaintance of an extraordinary range of scholarly gentlemen collectors, one of these being

Agassiz. It was Agassiz who had first introduced Sir Philip Egerton and Lord Cole (later 3rd Earl of Enniskillen) to the subject of fossil fish having met them while travelling. These leaders of British fossil ichthyology had been taught by Buckland, who then made use of Egerton's and Cole's collections.[40]

The purpose of Agassiz's trip to England was to examine collections of fossil fish for his ongoing work, *Recherches sur les Poissons Fossiles* (1833–44), in which he describes 1,700 species. In this great work Louis Agassiz acknowledged the help he had received from both Mary Anning and Elizabeth Philpot at Lyme Regis in correlating the fin spines with the appropriate teeth in various specimens of liassic fish. Mary was as familiar with liassic fish (fossil fish found in the Lias) as she was with reptiles.

Agassiz stayed with William Buckland at Oxford – one wonders what he made of the extraordinary set-up in Buckland's extensive apartments, with live animals in glass cases and others wandering around amid the stuffed animals and fossils.

However, it was noted that he and Buckland were a lot alike, remaining good friends after the visit, and he probably felt very much at home. After all, it was Agassiz who reputedly used the same kettle for boiling bones as he did to boil up water for tea and coffee, without first washing it.[41]

In October Agassiz went to Lyme to examine collections of fossil fish, especially the Philpot collection – thirty-four new (to Agassiz) species of fossil fish from the Lyme area alone – and to meet the Philpots and Mary Anning. He would view sixty collections on his travels, including those of his acquaintances Sir Philip Egerton and

Lord Cole. The Geological Society graciously made a
room available to him in which he could store the 2,000
specimens of fossil fish that he had borrowed from these
collections. Agassiz sent for his artist in France to come
and draw them – a project that took several years. When
he returned to Oxford he gave Buckland what he thought
was fresh and important news: the existence of specimens
with ink sacs in the Philpot collection.

Among the small fossils Mary had found were four
species of ammonite and *Ichthyodorulites* or fin spines of
Hybodus, a primeval shark with a large dorsal fin spine.
Mary recorded a discussion with Agassiz in her only
published writing, a letter of 1839 to Edward Charlesworth, editor of the *Magazine of Natural History*, on the
subject of the tooth of the *Hybodus*: 'In reply to your
request I beg to say that the hooked tooth is by no means
new; I believe M. De La Beche described it fifteen years
since in the Geological Transactions, I am not quite posi-
tive. But I know that I then discovered a specimen, with
about a hundred.' She continued that she had discussed
it with Agassiz; he remarked that they were the teeth by
which the fish seized its prey – 'milling it afterwards
with its palatal teeth'. Mary noted in her letter that she
was 'surprised' that he had not mentioned it in his work,
continuing 'We generally find the Ichthyodorulites with
them, as well as cartilaginous bones.'[42]

*   *   *

The 'discovery' of coprolites (bezoars) followed a similar
pattern to that of fossil ink. Like loose pebbles, these
innumerable oblong grey stones, usually 4 or 5 inches
long and with spiral markings, had been found at Lyme

for many years. Mary and the Philpots had recorded coprolites in 1824. Mary had noted that the bezoars were often found inside the ichthyosaurs, often in the pelvis, in the abdomen near the ribcage, or close to the creature, seemingly discharged at the moment of death. The markings on them remained from their passage through the intestines. In a letter of 21 December 1830 to Buckland Mary refers to coprolites: 'The bed containing the coprolites, belemnites, pebbles, bones, etc., I took Mr De La Beach [sic] to see and he was perfectly convinced that it was in the upper marles at a great height above the last bed of limestone (I should observed [sic] that the Coprolites etc rest on a conglomerate of shells about two or 3 inches thick undurated marls below the conglomerate of shells).'[43]

At a time when ichthyosaur bones were believed to be crocodile's teeth, and ammonites to be petrified serpents, coprolites were called locally 'Bezoar stones' because they looked somewhat like the 'concretions in the gallbladder of the Bezoar goat'. Mary and her mentor Buckland were both arriving at the same conclusion. Mary had put forward the theory that coprolites were the faeces of prehistoric reptiles, basing this conclusion on the coprolites she had seen inside the ichthyosaurs. Buckland agreed, basing his conclusion on the hyena faeces found in Kirkdale Cavern. In 1829 Buckland published the first paper on coprolites, in volume three of the *Transactions of the Geological Society.*

With Mantell, he concluded that they were found most frequently where there were numerous animal fossils. It was Buckland who named these small fossils 'coprolites' in 1828, but his research on the subject was very much tied in with Mary's. 'Our townswoman' was,

KIRKDALE CAVERN

Kirkdale Cavern, located about 25 miles north-north-east of York in the Vale of Pickering, was accidentally discovered by workmen in 1820. When Buckland heard about it, he rushed to investigate the cavern and its passageway recesses by candlelight, expecting to find evidence of The Flood. It must have been an astonishing and unnerving scene – there were hundreds and hundreds of bones of every description and from immensely varying periods of time.

In the 'bone cave', a hyena den where the animals returned with their prey, he found the bones of at least seventy-five individual hyenas, as well as still-existing species. But there was much more than that: prehistoric straight-tusked elephant and narrow-nosed rhinoceros, ox and deer, tigers and wolves. However, almost no bones were complete.

Billy, a spotted hyena imported from Africa, joined the menagerie of the irrepressible Buckland, and, to enable close study, was allowed to roam about the college apartments. He noted that Billy thoroughly gnawed and half ate the bones he was given; they looked identical to the cracked and chewed bones in the cavern. Similarly the fossilised hyena faeces he found in the cavern was just like Billy's. But Billy was from the tropics, larger and stronger than the native species, so Buckland correctly concluded that the cavern had once been tropical.

In 1821 Buckland made a detailed survey of the bones, and presented a paper on the cavern to the Royal Society in 1822. This work greatly enhanced his geological reputation, and he was awarded the Copely Medal, the first time it had been presented to a geologist. He wrote about his findings in *Reliquiae Diluvianae* (Relics from the Deluge). However, the bones were not water worn, and there was no evidence of The Flood. Buckland's belief in the literal story of the Creation in the Bible was shaken.

according to Roberts, '[Buckland's] handmaid of geological science'.[44]

Once the coprolites had been identified correctly it was possible to study what the creature had eaten before it had died. Coprolites contained an abundance of lime, the unmistakable rhomboid scales of the fish *Dapedium politum*, even undigested teeth, all of which could be studied. This meant one thing: the monstrous reptiles were eating fish and each other. A horrifying picture emerged of the stronger creatures devouring the weaker, and the violent death struggles – both fascinating and repellant – could only be imagined. To Buckland it was nothing less than 'warfare waged by successive generations of inhabitants of our planet on one another; and the general law of nature which bids all to eat and be eaten . . .'.[45] He was not at all restrained by what was considered to be a coarse subject. In fact he enjoyed being crude in his lectures. At Lyme he worked with Mary for weeks studying coprolites, and then travelled all over the country with his usual enthusiasm and obsessiveness in order to identify which coprolites were found in which rock formations. But Creationists were puzzled; they could not understand why God would have created these non-living things.

Toilet humour was found to be amusing, even when applied to prehistoric reptiles. An Oxford colleague, John Shute Duncan (1769–1844),[46] was moved to compose a poem:

> Approach, approach ingenuous youth,
> And learn this fundamental truth
> The noble science of Geology
> Is firmly bottomed on Coprology.[47]

*   *   *

In the early nineteenth century, the Lias at Lyme was quarried from the sea ledges by workmen breaking open the layers of rock. This material had an important use in London and other cities – to make stucco. When Mary sought out a particularly large beast, it was these quarry-men she called on to help. For smaller excavations, villagers and odd-job men assisted her. However, quarrying was destructive, exposing large areas of the Lias, accelerating its disintegration, and this also allowed the tides to wear it down even further, thus exposing more fossils. This left Mary with a search area, according to Roberts, 3 miles long and ⅛ mile wide.

Mary's next important discovery came in December 1828. It was a creature that is still the stuff of nightmares, a pterosaur: *Pterodactylus* ('winged-finger') *macronyx* ('big-clawed'), renamed *Dimorphodon* ('two-form tooth') *macronyx*. The early winter storms had dislodged a winged creature that has sometimes been called a 'flying dragon'. Pterosaurs were not dinosaurs but closely related reptiles. They probably lived along the shoreline, were lightweight with hollow bones, and may have been capable of powered flight. The wings consisted of a membrane supported by an elongated fourth digit. *Dimorphodon* had a wing span of about 4 feet, ate fish, and (as was eventually discovered) had a skull shaped something like the beak of a puffin. This was the first British specimen found, and is a rare find even today. Women were not accepted as Fellows of the Geological Society or even permitted to attend meetings, so it was left once again to Mary's old friend, benevolent, enthusi-astic William Buckland, to have the honour of describing this creature in lectures and in the *Transactions of the Geological Society* for 1829, read to the Fellows on 6 February 1829 – and to accept the glory: 'In the same blue lias formation at Lyme Regis, in which so many specimens of Ichthyosaurus and Plesiosaurus have been found by Miss Mary Anning, she has recently found the skeleton of an unknown species of that most rare and curious of all reptiles, the Pterodactyle, an extinct genus . . .'. A specimen had already been found in Germany, in the limestone beds of Aichstedt and Solenhofen (Bavaria), 'in the lithographic zone, which is nearly coeval' with the chalk of England, but nowhere else. Frightening in appearance, with small pointed teeth and bat-like wings, having three talons at the anterior joints, the *Dimorphodon* was, however, small – about the size of a raven with a 4-foot wingspan. Although the head was missing, the rest of the creature was there.[48] Such a strange and terrifying animal, seeming to combine the features of a bird, a lizard, an iguana and a bat, caught the public's imagination. Mary Anning became so famous that she herself was a sight to see in Lyme.

Once again, William Buckland had long suspected the existence of such a creature, but had never seen one. However, about twenty years earlier, he had heard of a specimen of a fossil 'bird' in the collection of a Mr Rowe at Charmouth. Additionally, in 1825, in the collection of the Misses Philpot, he had 'found' some bones of a wing and a toe, which he surmised must have come from such a creature. More recently, he had 'discovered' in *their* collection 'a thin elongated fragment of flat bone, which appears to be the jaw of a Pterodactyle; it is set with very minute, flat lancet-shaped teeth, bearing the character of a lacertine [Lacertilia, a sub-order of reptiles] animal'.[49]

Buckland was credited with Mary's discovery, and nothing could outdo his description in 'On the Discovery of a New Species of Pterodactyle in the Lias at Lyme Regis' in the *Transactions of the Geological Society of London* (1829):

> [It] somewhat resembled our modern bats and vampyres, but had its beak elongated like the bill of a woodcock, and armed with teeth like the snout of a crocodile; its vertebrae, ribs, pelvis, legs, and feet, resembled those of a lizard; its three anterior fingers terminated in long hooked claws, like that on the fore-finger of a bat; and over its body was a covering . . . of scaly armour like that of an iguana; in short, a monster resembling nothing that has ever been seen or heard-of upon earth, excepting the dragons of romance and heraldry.[50]

Mary Anning had unearthed the first British *Dimorphodon macronyx*. Some of the fossil can be found in the Natural History Museum, London,[51] but the teeth were in the Philpot collection and so they eventually went to Oxford.

*   *   *

In the emerging new sciences of geology and palaeontology, public curiosity was growing faster than national museums were being established. Until the Natural History Museum was completed in 1881, a rapidly expanding collection of specimens was held in the British Museum's Natural History Section. Also, entrepreneurs set up private natural history museums and exhibitions, especially in the large cities, which helped to satisfy the interest of the public.

One early museum was the 'Holophusikon' in Leicester House, Leicester Square, London, founded by Sir Ashton Lever, an obsessive who went bankrupt buying rare specimens. The result was that in 1785 lottery tickets were sold with the collection itself as the prize, and a land agent, James Parkinson, won. He kept it going for another twenty years, after which the collection was broken up and sold in a sixty-five-day sale, much of it to William Bullock (b. early 1780s, d. after 1843), who was from a family of showmen. Bullock's Museum of Natural Curiosities was located in the Egyptian Hall at 22 Piccadilly, London, until the once popular collections were auctioned off in 1819.

Visitors to Lyme increased as Mary Anning won the respect of contemporary scientists. In 1829, she found her *second* complete plesiosaur, this time a *Plesiosaurus dolichodeirus*. *The Dorset County Chronicle* reported the find on 19 February:

Miss Mary Anning of Lyme, has discovered another specimen of the Plesiosaurus Dolichoderius [*sic*] (a long-necked animal, almost a lizard). This specimen, which is 11 feet in length, is almost perfect, and most of the bones are lying in perfect order. The head, sternum, vertebrae, and bones of the pelvis and the paddles, are all fine and in place. There are four vertebrae between the last dorsal vertebrae and caroidal vertebrae, with their false ribs attached; by which it appears that this creature had the power of shifting its sternum, property of some amphibious animals now existing, extremely curious, but very useful when swimming. There also appeared between the ribs (which are exceedingly fine) a bony substance that must have been either a skin or a shell.

Mary Anning fossiling on the foreshore in the 1830s, an affectionate watercolour by her old friend Henry De La Beche, who enjoyed caricaturing those he knew. *(Reproduced with permission of Roderick Gordon and Diana Harman)*

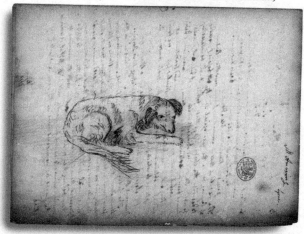

Mary's 1824 sketch of her constant companion, her faithful dog, Tray, whom she trained to guard a fossil find when she had to leave it temporarily. Tray was killed in a landslide before her eyes. *(©Natural History Museum, London)*

The elegant curves of the Cobb, Lyme Regis, today. The Cobb was cleverly constructed by the enterprising locals hundreds of years ago to create a harbour where there was none. This established Lyme as an important port. *(Author's collection)*

The steps known as 'Granny's teeth' are on the Cobb. Full of character and ancient looking, they date from as late as 1825. *(Author's collection)*

The House in which the
famous Mary Anning
lived when she first sold
fossils.
Sketched June 1842, by
W. H. Prideaux and
Edward Liddon
The round table for the
fossils used to stand in
front of the open cellar
window which was a work
shop.
Cockmoil Square

An 1842 sketch, believed to be of Mary Anning's home and shop until 1826 in Cockmoile Square. *(Lyme Regis Philpot Museum)*

Broad Street, *c.* 1850s, the second home of Mary and her mother from 1826. They had moved uphill, away from the threatening, unpredictable sea, to 'Top of the Town'. The building is just above the second cart. *(Lyme Regis Philpot Museum)*

The irrepressible and eccentric Professor William Buckland. *(©Natural History Museum, London)*

An even firmer good friend, Henry De La Beche. *(Geological Society/©Natural History Museum, London)*

The intellectual, Revd William D. Conybeare. *(Geological Society/ ©Natural History Museum, London)*

Mary's favourite, Roderick Impey Murchison. *(By permission of the Trustees of the British Museum)*

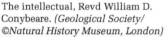

Always struggling, Gideon Algernon Mantell and his wife, Mary. (©*Natural History Museum, London*)

Thomas Hawkins, whom some thought mad, as he may well have been. (*Somerset County Museums Service/ Glastonbury Antiquarian Society*)

John Bennett, Mary Anning's neighbour. (*Lyme Regis Philpot Museum*)

Plesiosaurus macrocephalus, 1823, unusually, but handily for excavation, curled up in death. This was one of Mary's major finds. *(Geological Society/ ©Natural History Museum, London)*

The flying reptile *Dimorphodon macronyx*; it frightened and mesmerised all who saw it. With this important, still rare, discovery, Mary's fame spread far and wide. *(Original painting by John Sibbick/ www.johnsibbick.com)*

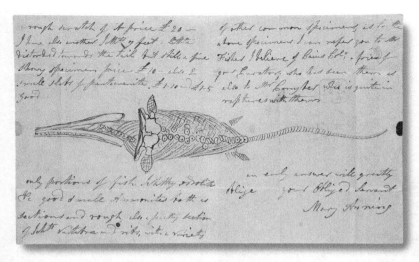

Mary's sketch, or 'scratch', as she would say, of an *Ichthyosaurus*, from her letter of 4 May 1843 to Adam Sedgwick, now in the Sedgwick Museum, Cambridge. *(By permission of the Syndics of Cambridge University Library)*

Cartoon by De La Beche of 'Professor Ichthyosaurus', based on Buckland, whose amusing lectures attracted both students and other lecturers. (*©Natural History Museum, London*)

AWFUL CHANGES.
MAN FOUND ONLY IN A FOSSIL STATE———REAPPEARANCE OF ICHTHYOSAURI.

A Lecture.—" You will at once perceive," continued PROFESSOR ICHTHYOSAURUS, "that the skull before us belonged to some of the lower order of animals; the teeth are very insignificant, the power of the jaws trifling, and altogether it seems wonderful how the creature could have procured food."

Mary Anning's grave at St Michael's, shared with her brother Joseph and some of his infants. The church is on a headland overlooking Lyme, the bone beds and the sea. *(Author's collection)*

Detail of the stained-glass window dedicated to kind Mary, appropriately celebrating the six corporeal acts of mercy. *(Author's collection)*

The picturesque Fossil Depot at the foot of Broad Street, run by a Sidney Curtis, *c.* 1900. This shop came after Mary's death, and was something of a misnomer, because the enterprising Curtis handily sold fresh fish as well as fossils! The whale bone, however, may have been one displayed by Mary Anning. *(Lyme Regis Philpot Museum)*

Various institutions were eager to acquire it. To overcome the dithering that had occurred regarding the acquisition of previous important specimens, Buckland immediately used his contacts and his irrepressible personality to insist that the British Museum purchase it, which it did for £100.[52]

*   *   *

Then, in December 1829, Mary made her fourth major discovery: *Squaloraja*, a new fossil fish. She first mentioned it in a letter of 15 December 1829 to Charles Lyell.[53] It was most unusual looking, unique, difficult to relate to any known creature, and Mary referred to it as 'nondescript'. The 1½-foot-long oddity was thought, wrongly, to be an intermediary between sharks and rays.

Mary wrote to Sedgwick on 11 February 1831 about 'the new fossil in my possession'. Part of her sales pitch to Sedgwick included a sketch, which she calls a 'scratch'. She did her best to describe it as 'a skeleton with a head like a pair of scissors Vertebrae like an encrenite [encrinite] thin as a thread of which there are two 100 & 52 and the tail wanting the greater portion of six claws or felers [feelers] . . . the vertebrae skin and snout covered with tubercles like those of the ray tribe which it strongly resembles in some parts and wholly differs in others . . . being the only one in Europe 50£'.[54] Sedgwick did not acquire it, but a wealthy landowner, John Naish Sanders, did. He purchased it for the Bristol Institution, paying about £40. Although Mary promised to send the tail when she found it, the Philpots found it first, so this appendage entered the Philpot Collection, and therefore later went to University Museum, Oxford, where it remains.[55]

Dr Henry Riley (1797–1848) first described it in 1833 in a paper read to the Geological Society, but the creature was not fully understood until much later. Riley had a fine collection of fossil fishes and was a skilled doctor, although in 1828 he and another doctor had been arrested for grave-robbing, but had got off with a fine of £6 (about £200 today) each.[56] It was finally identified in 1881 as a chimaera, a fish of the family *Chimaeridae*; the male has a frontal spine and a long thin clasping snout over the mouth. It came to be known as *Squaloraja polyspondyla* (Agassiz would illustrate it in *Poissons Fossiles*), and, although incomplete, Mary's find would become the type specimen.

*   *   *

Another major discovery came in December 1830. It was Mary's fifth. She wrote to Buckland with respect, even affection, on 21 December 1830 announcing her discovery of another species of plesiosaur, *P. macrocephalus* (large-headed):

I write to inform you that in the last week I discovered a young Plesiosaurus about half the size of the one the Duke had, it is without exception the most beautiful fossil I have ever seen. The tail and one paddle is wanting (which I hope to get at the first rough sea) every bone in place, in short if it had been made of wax it could not be more beautiful, but . . . the head is twice as large in proportion as those I have hitherto found. The neck has a most graceful curve and what makes it still more interesting is that resting on the bones of the pelvis is, its Coprolite [fossilised faeces] finely illustrated.[57]

This large-headed reptilian creature, preserved curled up in death, was purchased by William Willoughby Cole, Lord Cole, for 200 guineas (about £6,800 today). Named by Buckland, it was described by Richard Owen in 'A description of a specimen of the *Plesiosaurus macrocephalus*, Conybeare in the collection of Viscount Cole', 1840.[58] An even larger ichthyosaur was excavated by Mary in 1832: *Temnodontosaurus platyodon*, Conybeare. On view in the Natural History Museum, it measures 6.83 metres (22½ feet). This is one of the largest specimens ever found, and today retains the power to awe those who look upon it.[59]

Mary explained to writer John Murray how she could estimate the length of a creature by looking at the dorsal vertebrae. A vertebrae 1 inch in diameter was equivalent to 5 feet in the length of the creature.[60]

Although it was Mary Anning's discovery of the large fossil vertebrates that brought her fame, she also found many smaller specimens, including innumerable ammonites and brittle star (*Ophioderma egertoni*), which is similar to a starfish. To collect the larger ones she needed to use all her ingenuity and skill, taking the finds home in slabs and working on them there to preserve them. Throughout it all she was imbibing information, always adding to her experience and knowledge, and building up a database in her mind. It was she – not the learned gentlemen – who was on the foreshore day after day, year after year, sensitive to each bump and depression and what it might be.

Mary was 30, and she had already made five major palaeontological discoveries. She did not know it, but the golden years – the succession of extraordinary discoveries by the carpenter's daughter, the fascination for fossils, and the high prices paid – were coming to an end.

4

Reptiles and Relationships

Look at the bones. Did you ever see a more elegant, a more stately, a more severely geometric set of abstract lines? Dark against dark; clean lines; primary shapes . . . Thomas Hawkins' formless insanity secured these specimens; Mary Anning's stately order prevails.

(Stephen Jay Gould[1])

She was a woman in a man's world, and the men in the main were eccentric, some wildly so. Mary Anning herself would have been considered by many to be quirkish, even freakish. In a highly sexist society, a spinster, poor and in trade, and in such an unusual trade, was someone to be pitied. But her eccentricity sprang from the need to put food on the table, not as the hobby of a gentleman, among a group who were either amateur fossil collectors or professional geologists with the stature, money and time fully to indulge whatever interests and obsessions popped into their heads.

Her kindness to children and the sick was no doubt genuine, and also made this rather odd woman more acceptable to those she lived among. By far her greatest contribution to the community was her own fame, which attracted a succession of prominent scholars, visitors and money to Lyme over many years.

She seems to have enjoyed a solitary life, yet the arrival of these eccentric men with their personalities, news and gossip must have been very welcome and enlivening in quiet Lyme. They ranged from the wonderfully unique and lovable William Buckland – his laughter ringing out as the pair of them waded knee-deep along the foreshore – to a more strange and difficult example of human nature: Thomas Hawkins (1810–89). In both cases the individual had the energy to match the obsession. And after they had left she could relish her quiet, private life and thoughts, as she explored 'her' terrain.

One might think that it would be hard to outdo Buckland in bizarre behaviour, yet Thomas Hawkins managed to do so. Unlike Buckland, his personality was at times strange and often confrontational. Some thought him mad. Described by J. Clark as being 'of middle height, light hair, and a foxy unpleasant face',[2] Hawkins was described by Lang as one 'who did nothing by halves and moved in an atmosphere of superlatives'.[3] From about 1832 to 1845 Hawkins lived at Sharpton Park near Glastonbury, once the home of Henry Fielding, author of *Tom Jones*. It was close to the richly fossiliferous Lias quarries at Street, between Bristol and Lyme, from which he excavated many fossils. Sir Richard Owen described breakfast during a visit in 1839: 'That worthy and eccentric man of genius had procured me peacock's eggs . . . no bad things by the way – and other rarities conformable . . .'.[4]

However, Hawkins had a perverse gift for converting almost everything into exceptional drama. The provocative fossil collector in reality only rented the house – in fact just part of it – and he was soon locked in a violent dispute with the unfortunate owner. This escalated into nothing less than open warfare between the nearby villagers, who sided with the owner, and a gang that Hawkins employed to guard 'his' house. It is but one example of where the unbalanced, litigious side of his personality led him. His accounts of his own life were unreliable. No doubt seeking to enhance his social standing, he made claims ranging from being the 'Rightful Earl of Kent' to having saved Robert Peel from assassination. As was to be expected with a personality such as his, Hawkins collected on an enormous scale in order to find these 'wrecks of an often wrecked world,'[5] causing normally calm and composed De La Beche to exclaim, 'Why, you carry away whole quarries'.[6]

Mary Anning was involved in at least two of the excavations, which he described in his first book, *Memoirs of Ichthyosauri and Plesiosauri* (1834). This oversized volume was written in his usual excessively florid style and published (could it have been otherwise?) in an expensive imperial folio volume. The list of subscribers included Miss Philpot, but not Mary Anning; it would have been too expensive at £2 10*s* (almost £95 today). He also wrote *The Book of the Great Sea Dragons* in 1840. With his typical fervour, Hawkins gave an account of his visit to Lyme and the extraction of his first ichthyosaur. He certainly had a good sense of timing, for on the day he arrived in July 1832, 'Miss Anning obtained from the indurated marl of the lias limestones near St Michael's, the parish church at Lyme, part of the head of the Chiroligostinus [Hawkins's name for *Ichthyosaurus* – he persisted in giving the creatures his own Latin names based on the shape of their limbs] . . . Happening to arrive at Lyme the same day, I was fortunate in availing myself of the specimen.' The massive head was 5 feet long, and the creature had, untypically, come to rest on its stomach. Having sold him the squashed skull, the next morning Mary took him to the foreshore and pointed out where she had found it. She assumed that no one would be able either to find or to retrieve the rest of the enormous beast, which was hidden and securely buried in a protrusion of the cliff.

Hawkins had the head, and now he was determined to possess the rest of it, 'but Miss Anning had so little faith in my opinion, that she assured me I was at liberty to examine its propriety or otherwise myself'. However, he was able to rally more resources than Mary could have dreamt of. A major earth-moving exercise was soon under way to get to the treasure quickly. First he sought and received permission from the owner of the land, Mr Edwards, 'to throw down as much of the cliff as was necessary', then he hired as many men as it took to carry out the work.

'The sun rose bright' on 26 July when these men attacked the site with spades and pickaxes for two and a half days while people flocked to watch the activity. Twenty thousand loads of earth were dislodged, used to make a roadway to the beach, and finally the 'wonderful remain' was revealed. '*My* eyes the first which beheld it! – who shall ever see them lit up with the same unmitigated enthusiasm again!' It would prove to be the largest *Ichthyosaurus communis* found to date.

Hawkins described the fate of the cliff, outlining what it had been like before:

four or five hundred yards of the coast from the
Borough Eastward has an elevation of from sixty to a
hundred feet above high-water mark, and that a bed of
diluvial gravel conceals the blue-marl of the lias from
observation, except in those places where the rain has
ploughed itself a channel towards the sea. It was at the
place where was seen . . . a kind of peninsular rock,
which had long defied the fury of the destructive cur-
rent that a South-Wester invariably propelled against it
from the cob. There it abutted upon the angry waves,
reft of its gravelly covering by the storm, with its grey
sides slowly crumbling beneath the frost and saline
atmosphere; but its foundations sound and unmoved.

Having been there from time immemorial, it could not
withstand Hawkins's onslaught; but then, nothing could:
'that venerable though tiny promontory is no more. What
the warring elements failed in, curiosity achieves; the
hand of man came upon it, and it departed like a
shadow.' In the tug-of-war to get them out, the bones,
with the marl in which they lay, broke into small pieces –
there were 600 of them. 'With the kind assistance of Miss
Anning, the whole of them were packed, and by night-
fall the last heavy box-full was deposited in a place of
safety. So secured the skeleton and its matrix weighed a
ton.'[7] It was subsequently acquired by the Trustees of the
British Museum and is now exhibited in the Natural
History Museum.

The next year, 1833, Hawkins obtained another
ichthyosaur from the foreshore between Lyme and
Charmouth. He left a long-winded, effusive account of
how he obtained it. Others, besides Mary, were searching
for fossils tosell. In August Hawkins called on the fossil

collector Jonas Wishcombe at Charmouth to see if he had found anything worth buying. Indeed, something had been found.

'Have ye sid my animal sir,' Wishcombe asked. 'I should like vor yer honour sir to see 'im . . .'

Hawkins later wrote, 'My heart leaped to my lips – animal! animal! Where!'

'Can't be sid today – sir – the tide's in.'

'What – nonsense! I *must* instantly – come, come along,' retorted Hawkins.

'Can't see 'un now yer honour – the tide's rolling atop'o'um fifty feet high.'

The next day they were on the beach at extreme low water. Hawkins recorded: 'The weather-cock looked the right way as we descended the execrable path by which the good people of Lyme are content to wade to the sea-shore between their delightful town and the pleasant village of Charmouth.'

Wishcombe spoke to Hawkins: 'Do ye see that tuff o' sea-weed just tipping up there yer honour; he's there about.' The keen seller of a prehistoric creature waded out and swished the wet sand off the entombed 6-foot *Ichthyosaurus communis* with a furze bush he carried with him for that purpose.

Then the 'adored lizard' was at Hawkins's feet. Hawkins waxed lyrical when it came to actually laying eyes on what he could see of the specimen:

Let every naturalist make pilgrimage to the beetling crags mid-way of the rivers of Char and Lim. A waterfall issuing from a placid lake in their lone bosom bounds from ledge to stony ledge and is swallowed up of the thirsty shingle at their base . . . there lay the

Chiropolyostinus [*Ichthyosaurus communis*] the mark
of the vertebral column of which was scarcely percept-
ible through the selenitic layer that had preserved it
from the roaring waters.

Wishcombe had been quietly savouring his find for some
time, and now smugly sold the rights to Hawkins. He was
even more delighted when Hawkins gave him an extra
guinea in 'earnest money'. Wishcombe believed he had
executed a clever deal in selling something that could not
be extracted. He was not the only one.

Mary Anning, her woven, rush collecting bag over her
right arm, clutched her cloak around her with the other.

'You will never get that animal,' Mary said, as they
made their way towards Lyme through the mist and
flashing spray, 'or if you do, *perchance*, it cannot be
saved . . .' She saw Hawkins's 'change of blood', and
stopped to look at him.

Hawkins noted, 'My eyes glare upon the intellectual
countenance before me – the words of those lips were I
knew those of a Pythoness [Pythia at Delphi, Apollo's
priestess of the oracle] and my heart fainted within me.'

'Because the marl, full of pyrites [iron pyrites or 'fool's
gold'], falls to pieces as soon as dry,' Mary added.

'That I can prevent,' said Hawkins.

'Can you?'

One can imagine Mary's sardonic look accompanying
this cool reply.

Always in a hurry, this dynamo was, however, compelled to wait 'some weeks' while the wind blew from
the south-west. Ever more impatient, day after day
Hawkins watched the 'ugly fish' weather-vane set on the
'ugly tower' in the 'ugly marketplace' of Lyme. At last the

wind changed, the weather brightened, and it was possible to try again. The site was dry only when there was a strong offshore breeze, preferably with an equinoctial tide.

'Make haste; the tide's going out fast,' warned Mary, as she and Hawkins passed each other on the foreshore.

He took this encouraging comment from Mary – 'the presiding Deity', as Mantell had called her – as a sign of good fortune: 'I seized this opportunity of thanking her for this brief exhortation; it secured me the saurus [lizard] the same day. Really the tide seemed to gallop away. Half a dozen of us – all lusty and eager for the occasion – meet.' They discussed how they would exhume the giant specimen that was sleeping in its 'oozy bed', got their tools ready and waited for the retreating waves to expose the remains. They marked out a square around the creature 6½ by 3 feet, within which only Hawkins would excavate. Then it was time to launch into the great task; crowbars and pickaxes loosened the creature in the square, but the men were not strong enough to turn it on its side. Then even those spectators representing the nobility and the church, who had had no intention of getting their hands dirty, were ensnared by Hawkins's maddened enthusiasm. Sir Henry Baker and the Revd Benjamin Jeanes joined the sweating, toiling labourers in the endeavour. It was a close-run thing, but the specimen was retrieved.[8] As Mary Anning had predicted, when the marl dried it cracked. But Hawkins was ready for this too, and had carpenters on hand to enclose the remains of the 25-foot creature in a tight case of immovable plaster of Paris. A print of it was made by Dunster of Lyme, without the missing fore-paddle, that is, before Hawkins's full 'restoration', and this became an insert in *Roberts's History of Lyme Regis and Charmouth* (1834).

Whatever Hawkins's faults, he did not stint in giving fulsome credit to Mary Anning in his *Memoirs*, and also acknowledged her valuable assistance to Conybeare, Buckland, De La Beche and others:

> it must never be forgotten how much the exertions of Miss Anning of Lyme, contributed to assist them. This lady, devoting herself to science, explored the frowing and precipitous cliffs there, when the furious spring-tide conspired with the howling tempest to overthrow them, and rescued from the gaping ocean, sometimes at the peril of her life, the few specimens which originated all the fact and ingenious theories of those persons, whose names must be ever remembered with sentiments of liveliest gratitude.[9]

Hawkins was famous for over-restoring his specimens. Once in his possession, an almost whole animal – or even a somewhat incomplete one – miraculously became beautifully perfect. In Mary's opinion, 'if Mr Hawkins has set the last specimen of Icht platyodon as it lay in the cliff, it will be a most magnificent specimen, but he is shuch [*sic*] an enthusiast that he make things as he imagines they ought to be; and not as they are really found, the platyodon that I in part gave him was to large for my poverty & I would not have trusted to his making up, though very much broken, it might be made a splendid thing without any addition'.[10]

After complicated negotiations involving Buckland and others to smooth the way – with Hawkins as pup-peteer manipulating all concerned behind the scenes –

Hawkins sold two collections to the British Museum in 1834 and 1840 for £3,110 5s (today about £118,000). In these unique collections, deemed to be the finest ever assembled, were superb examples of ichthyosaurs and plesiosaurs, some collected by Mary Anning. However, later the rather awkward question arose of just how much restoration had taken place. When closely examined, the specimens were found to be very much over-restored, placing Buckland in a difficult position, since he had vouched for them after only a cursory examination. This dispute led to an investigation by a Select Committee of the House of Commons. Hawkins's collections, along with the 1838 acquisition of Mantell's hoard of fossils – 20,000 specimens assembled over twenty-five years – for £4,087 (today about £133,000), made the British Museum's collection of fossil reptiles the most important in the country (now in the Natural History Museum).

One of many examples of Hawkins's strangely overt behaviour was the publication, in 1844, of his epic *The Wars of Jehovah in Heaven Earth and Hell*. Reflecting his overstated personality, it appeared in imperial quarto.

Among these great characters, De La Beche must have seemed an ever constant island of sanity to Mary. Although he had some eccentricities himself, this was balanced by a sense of humour, which he expressed in both verse and caricature. Even *his* personal life was not entirely tranquil. His wife left him in 1825, after seven years of marriage and a child, due, she said, to his behaviour; but she had taken a lover. He, ever the gentleman, took the blame. In 1819, he had been elected a Fellow of the Royal Society, and President, in 1848–9, of the Geological Society, and remained an active member for the rest of his life. He travelled to a number of sites of geological interest, both in Britain and in Europe. In the south-west of England he began the detailed examination of the rocks of Devon and Cornwall. De La Beche's accomplishments were many. He published numerous memoirs on English geology in the *Transactions of the Geological Society of London*, as well as in the *Memoirs of the Geological Survey*, which he founded, notably the Report on the Geology of Cornwall, Devon and West Somerset (1839). He likewise wrote *Researches in Theoretical Geology* (1834), in which he enunciated a philosophical treatment of geological questions in advance of his time, and much else.

*   *   *

De La Beche and Mantell were two of the main characters in Mary's story. Mantell, with all his obsessive, restless energy, had great success, yet in some ways was unlucky. He moved to Brighton in 1833 – seeking more good fortune both as a physician and as a fossil collector in the seaside resort where George IV had his Marine Pavilion. But everything that could go wrong, did go wrong. In Brighton he had little business either as a doctor or as a fossil collector/geologist. The house became a museum, driving his family out. Then he bought a medical practice in Clapham in search of financial stability, and, in 1838, sold his entire collection to the British Museum, as already noted. In the end he had lost his collection and most of his family, but spent the next ten years working creatively, despite being very ill.

For twenty years Gideon Mantell had been the leading light in the 'Age of Reptiles' with his discoveries of *Iguanodon* and *Hylaeosaurus*. There were also many other firsts – he was one of the discoverers of *Megalo-saurus*, which Buckland would identify. One of his great finds was *Pelorosaurus*, the first branchiosaurid dinosaur found. There was much more, including several dozen vertebrate and invertebrate fossils.

But, inexorably, time and the spotlight moved on to someone else: Richard Owen (1804–92), later Sir. Owen, a tall, thin man with 'glittering' eyes, had many outstanding qualities as a comparative anatomist and palaeontologist. Gifted at interpreting fossils, he was open-minded and respected. It was he who would be regarded as Cuvier's successor in vertebrate palaeontology under the patronage of Buckland. However, the admirable qualities of the foremost natural scientist of his generation were offset by a number of less desirable features.

He was ruthless, intellectually arrogant, obstinate and mistrusted. Owen and Mantell, once friends, had long since become bitter enemies, immersed in the squabbles of scholars and collectors. Each enjoyed overturning a discovery of the other with another new find. When he heard that the Royal Society intended to award the Gold Medal to Mantell, Owen – never a pleasant man – tried unsuccessfully to stop it. Attacks on Mantell added to the fear and hatred with which Owen was regarded by many of his contemporaries. However, it was largely due to his strenuous efforts and determined personality that the Natural History Museum was completed and opened in 1881.

Controversies and jealousies erupted frequently among the pioneer geologists. Friends sparred with each other, fell out, made up or became lifelong enemies. De La Beche's work on Devon was attacked by Buckland, Conybeare and others, almost destroying his career, and

Conybeare and Lyell fell out, or at least became friendly antagonists, regarding the still thorny subject of Creation. Buckland had furtively tried to use Mantell's work on the *Megalosaurus*, and so it went on.

Mary Anning first met Richard Owen in 1839, when he visited Lyme Regis. Owen will forever be remembered for naming the prehistoric reptiles *Dinosauria* (literally 'terrible lizards') in 1842. Owen went to Lyme in September 1839 to make notes for his monograph on fossil reptiles. He wrote a light-hearted letter to William Clift, the conservator/curator at Glasgow's Hunterian Museum, named after John Hunter; initially, Owen was Cliff's assistant. In this letter, written on the 13th, Owen referred mockingly to the mannish spinster he had met there: 'I may spend a day with Mr Hawkins at Street, then take a run down to make love to Mary Anning at Lyme, then post home.'[11] It is unlikely that he would have referred to Elizabeth Philpot in this way.

It is easy to believe that Mary was at times an object of ridicule – to Lymeites as well as others. When the Pinneys first met Mary, Anna Maria's mother insisted that a friend accompany her daughter 'as she was afraid to trust me alone with this extraordinary character, why I do not know'.[12] The locals and visitors of her day perhaps appreciated this gifted woman most after her death, when their living tourist attraction was gone for ever.

After visiting Hawkins, Owen wrote to Clift, 'From Sharpham I went to Lyme Regis, and there I met Buckland and Conybeare. They made me prisoner, and drove me off to Axminster, of which Conybeare is the rector. Next day we had a geological excursion with Mary Anning, and like to have been swamped by the tide. We were cut off from rounding a point, and had to scramble over the cliffs. I spent the next day in Miss Philpott's [*sic*] Museum; then went to Charmouth, and so returned to London.'[13]

Although Mary Anning was sometimes regarded as a figure of fun to the learned men with whom she consorted, she was later remembered for her interest in and kindly care of the sick, one of the corporeal acts of mercy, based on the teachings of the Church taken from the Gospel of St Matthew 25: 34–46. And, true to form, in her letter to Frances Bell she wrote, 'I had nearly forgotten to tell you that, three weeks back, poor Mrs England got upon a chair, to wind up the jack, and was seized with a paralytic fit, and fell backwards on the stone kitchen, which deprived her of all appearance of life for two days; and for a week her life was despaired of.' Conscious perhaps of her lack of social standing, Mary concludes her letter with niceties, apologies and careful assurance that there is no lack of respect. 'You must excuse my blunders, for I have neither time nor spirits to look this over; but be assured it is not out of disrespect: I retain too grateful a sense of all your kindness to me for that to be the case. Mother joins me in respectful regards to you, and your dear mamma; and aunt, when you write to her. God bless you! And may his choicest blessings rest on you! Is the prayer of . . . Your sincere friend, Mary Anning.'[14]

Outlets for Mary's emotional life, which normally would have focused on a husband and children, were her mother, her dog Tray, children and the sick. In a strange episode, when Mary was just 16, the body of a beautiful woman was washed up on the shore near Lyme, where Mary found it. Anna Maria Pinney recorded what happened in her journal for 1832. Although the event had happened some time before, on 27 March 1815, it was obviously still discussed. Mary had untangled the seaweed from the drowned woman's long hair, and performed all the necessary rites of preparing a body for burial. It was placed in the church awaiting someone to claim it, and every day Mary went there to strew fresh flowers over it. She was dutifully observing an act of mercy. But excitable Miss Pinney saw it in a more extravagant light: it was an example of the 'wild romance' in Mary Anning's character. The deceased woman turned out to be Lady Jackson, returning from Bombay with all her children on an East Indiaman, the *Alexander*, which was lost off Portland. There had been 160 people on board; only six lived to tell the tale.[15]

The Great Storm of 1824, which wreaked havoc on Lyme itself, was only the latest in a succession of violent tempests that were an ever-present threat, as they always had been. In 1407 a plea had been addressed to Henry IV by the Mayor and Burgesses of Lyme requesting remission of the dues owed him 'by reason of the sea's rages, as also by enemy attacks, our houses burnt, the plague and other destructive things' – one of many pleas to a monarch over the centuries.[16] And there was still a folk memory of a terrible storm on 11 November 1377, when the Cobb had been swept away completely, as was every ship that had sought shelter in it.

In Mary's lifetime, the Annings' house was in a dangerously exposed position. According to an account by Miss Marigold Watney, when Mary was a small child the Annings' home was in great peril during a violent storm. That night the sea hurled debris against the front wall of the house, water rushed in, and, as it pulled back, the staircase was washed away. The Annings were rescued from an upper window when morning light came: 'the Anning family . . . found that the ground floor of their home had been washed away during the night. Their house was in Bridge Street, that curious little row of buildings which spanned the mouth of the river Lym, and an exceptionally rough sea had worked the havoc.'[17]

On 20 January 1817 a storm had destroyed part of the Cobb and wrecked several ships. But the impact of the November 1824 storm was even greater. A freak high tide of 23 feet and a hurricane from the south-west combined to create a major disaster for Lyme. The never-to-be-forgotten turbulence severely damaged the Cobb, and the houses nearby were either wrecked or flooded.[18] Fanny replied to Mary's letter of 26 November, writing by candlelight on 18 December to 'My very dear Mary'. In this letter she notes her continuing faith in God in all her afflictions and 'the crisis at which my arm is arrived'. The doctor has assured her that whatever happens is for the best. She tells Mary that her misfortunes are lightened by thinking of what happens to others, for example the storm at Lyme, and tries to console her friend, 'and you, I doubt not, in contemplating the super-lative miseries of your fellow sufferers, are thankful that *your own are not so bad*'. She continued, 'Dear beautiful Lyme, what alterations shall I see when I visit you again!', and begged Mary to tell her more about the storm and what had happened to her own house, whether she had lost or found any fossils because of the storm, and more about the Roberts and 'my dear little cat'.

Frances Bell praised Mary for all she had taught her about fossils, and in a playful reproach added that she could not teach Mary anything regarding her own visits to the British Museum. She concluded by wishing Mary and her mother 'many more happy returns of this cheerful season'. But it was to be Frances's last Christmas. She was not well enough to make the visits to the museum as she had intended. By the end of May 1825 she was dead, aged 15½. The disease to which she succumbed is not recorded.

Two more letters survive from Mary Anning relating to this singular friendship. On 10 August 1825, Mary wrote to Frances's aunt:

> The young man delivered the parcel quite safe: my dear lamented little friend's bequest will always be revered by me. Dear kind-hearted child! Admidst all her sufferings to have thought of me and my comfort: as to my remembering her, never, whilst life remains, can I forget the transient vision of her friendship. Oh! madam, had you heard her kind pious conversations with me when we were alone, you would say that I was the most ungrateful of beings if I ever forgot her. Although so young, her mind was so heavenly gifted, that not to be doing good was to her impossible: and I trust that, *the trials which in this world I am doomed to encounter*, I shall think on her pious example, and submit without a murmur to the decrees of Providence; convinced that he only afflicts for wise purposes . . .

The second letter was written on 19 October 1826 to Frances's mother. These letters have survived because Mary kept them close by her, and because Frances's family had asked for them back so they could be reproduced in a Memoir. Mary was happy to be

> able in any way to gratify you; and much as I value the letters of my lamented little angel friend, if they afford

you a melancholy pleasure, pray keep them their contents are indelibly stamped on my memory; indeed I have reason to bless the day when I first saw dear lamented Frances. Although so many years younger than myself, she was my spiritual guide: the recollection of her pious conversations has been *a support to me in the trials I have had to sustain*; it has enabled me to say, 'Not my will, but thine, O Lord, be done.' My dear madam, we must endeavour to look on this world as a state of trial, to fit us for a better life.

She added, 'I never think of my lamented friend but my heart is full.'[19]

These exchanges, while effusive today, reveal a caring, motherly side to Mary Anning. And equally important, this early fossil collector whose discoveries over and over again disturbingly challenged the religious status quo, was herself a deeply pious person.

*   *   *

There were other friends. Roderick Impey Murchison (later Sir) and his wife, Charlotte, had first met Mary Anning the year after the plesiosaur find, that is in 1825. She seemed to have an easy-going relationship with them both, in spite of the class difference, and Mrs Murchison often acted as an unofficial agent for Mary, providing her with fossil-finding commissions. Mary would have seen through any insincerity instantly.

Murchison – tall, muscular and wiry – had been an army man, until his services were no longer required after Waterloo. But he never lost his commanding presence and dignity, which was charmingly combined with courtesy and kindliness. His fitness was enhanced by his extraordinarily long walks: he once walked 452 miles in fifteen days. Charlotte Murchison was cultivated, intelligent, and could converse with a range of people on a great many subjects. Sedgwick later said that she was one of the dearest of those friends whose society formed the best charms of his life.

They became good friends, and Mary was invited to visit London in early spring 1829, staying with them at their house, 3 Bryanstone Street, south of Regent's Park. If Mary had visited after 1839–40, by which time Charlotte had inherited a fortune, she would have found the Murchisons in a mansion at 16 Belgrave Square, near Hyde Park Corner, where they welcomed men of science, literature and the arts at their soirées. John Ruskin attended a reception there, one of several hundred guests: 'I don't know what fortune Murchison has – but this is coming it rather strong – rooms all pale grey and gold – magnificent cornices – with arabesques like those of Pompeii in colour, furniture all dark crimson damask silk and gold – no wood visible – at least four footmen playing shuttlecock with peoples names up the stairs.'[20]

Mary Anning looked forward to the forthcoming expedition with the nervous, hopeful anticipation of one who has never made an overnight journey away from home, and now she was invited to visit the largest, most dynamic city in the world. She commented, 'should anything occur to prevent my accepting it, it will be the death of me'.[21] Something did occur. Another communication to the Murchisons followed at the end of February 1829. Mary explained that a letter necessary for her journey to London had not arrived in time, nor was it likely to. She was very disappointed, noting the honour she felt it was to be invited to their home, and adding resignedly, 'I have never been out of the smoke of Lyme.' She continued, 'I can truly say that hope deffered [sic] maketh the heart sick.' In a footnote she sent her regards to Mr Murchison, adding 'and tell him I long to examine his cabinet more especially to examine his bones'.[22] Although, she refers to 'the plesiosaur sent to paris' (she found the first plesiosaur in 1824), the paper is watermarked 1829, and the visit was probably made from 7 to 12 July 1829.[23]

She finally made what seems to have been her only visit to the great metropolis, probably as a guest of the Murchisons, although there is no certainty where she stayed, and the exact date is unknown. Did her friends, the Philpot sisters, or Mrs Murchison advise Mary on what to wear on her excursion to the great city? It is hard to imagine that they dared. Having grown up a dissenter, Mary had been taught that any excess in dress was offensive. She would have made her own clothes; all clothing was hand-made, and fabric was precious; when a garment was worn out, good sections of material were reused for other purposes. To a dissenter, ribbons and bows had to have a purpose, and decoration was not a purpose. The fashion – for upper-class women – was to emphasise the hips, buttocks and breasts by pulling in the waist. The physically restraining corsets and crinolines made an overtly public statement that work was not for them. Even breathing might be difficult. This public display would have been anathema to dissenters. Working-class women wore simple, loose clothing, that was both functional and hard wearing, no matter what they aspired to. When delicate paleness of skin indicated wealth and upper class, Mary looked tanned, strong and healthy. She saw so-called fashionable people all the time in Lyme and was not impressed, certainly not with 'the young dandies', whom she laughed at 'extremely'.[24]

Although Lyme was bathed in fresh sea air, Mary was accustomed to the smoke from the coal fires in every house, which managed to pollute the air and cling to the valley floor. (Coal was the most important commodity to come in to the Cobb by ship at that time.) But the smoke of Lyme, with its population of 3,345 (1831 census), would not have prepared her for 'smoky London', with a population of almost 1,700,000, which her 'dear lamented little friend', the invalid Miss Bell, had told her about. Mary probably travelled to London by boat; it was possible to go by sea in a vessel of about 80 tons, departing from the Cobb every other Saturday. There were three ships offering this service, one being the *Unity* – following the same route on which she sent her fossils on their journeys to the individuals and large institutions she supplied.[25] On reaching the busy river highway of the Thames, she would have passed mansions with gardens down to the river, warehouses bustling with activity, the Tower of London, the Custom House, then Billingsgate Market, disembarking in the heart of the city just below London Bridge.

Assuming the year was 1829, old London Bridge was still one of the sights to see, although long bereft of houses. Approaching the final year of its 622-year history, the bridge would have been teeming with hundreds of people, carriages and animals either going in or coming out of London, all adhering to the 'left-hand drive rule' that originated there. Workmen clambered over John Rennie's emerging new London Bridge, in its last stages of construction next to the old stone bridge; bales of hay hanging underneath warned Thames boatmen of the activity under way above their heads. Buckland had been on the committee to select the best granite for the new bridge.

Did Mary see London's gaslights, glittering and magical, but with the glow smudged by the smog, and the streets where there was little light and crime flourished? Only in 1829, just before Mary visited, had Sir Robert Peel got the first police bill through Parliament. In 1848 reforms included the Bow Street Runners, a separate force, instigated by Henry Fielding – the very same Fielding who, in his intemperate youth, famously attempted a kidnap in Lyme. Could Mary sleep in the teeming metropolis, the hub of the empire, the largest port and city in the world? With the strange, often frightening noises of the city ever present, she would have missed the pounding of the waves on Lyme's shore, a rhythm that had been a second heartbeat to her since birth. During the day, did she feel dizzy with the noise, claustrophobic with the press of people of every description, or did excitement overrule all else?

Mary left a portion of a diary of the London trip,[26] transcribed in her own hand and, as usual, with little punctuation and her own spellings. She began with the Geological Society, in Somerset House, Strand. It had been constructed early in the nineteenth century around a large courtyard, the first purpose-built office block in London – even though it looked like a palace. Esteemed institutions such as the Navy and the Stamp Office were housed there, as well as several learned societies. Naturally, the place where her own major finds had been announced was of primary interest to her. This was the address to which her plesiosaur – the first almost complete *Plesiosaurus giganteus* – had been delivered in 1824 to intense interest. She could imagine the flustered scene in the stairwell, when it was found to be too large to get up the stairs. She described the important visit in her diary:

Tuesday went to the Geological Society's rooms, Somerset House [north wing], the Teritary [Tertiary] fish very much resembles ours Mr Lonsdale more than kind showed and explained every thing particularly the gradations between the Mastodon, Rhinoceros and Elephant and that the Elephants Molar teeth when young were mammillated [mamillated meaning having spherical breast-like protuberances] at the grinding surface as they Grew older they were worn down to a flat surface, saw a model of a jaw of a large Megliosaurus [*sic*] also of the plesiosaurus sent to paris so like that I could hardly distinguish the difference . . .

The 'Lonsdale' Mary mentions was William Lonsdale (1794–1871), a geologist who made the first comprehensive study of the geology of the Bath area, and did much else of value. Murchison memorably described Lonsdale as a 'tall, grave man with a huge hammer on his shoulder'.[27] As well as work in the field, he was noted for his industry in preparing, labelling and cataloguing specimens in Bath. He was elected a Fellow of the Geological Society in London in June 1829, and on 1 June of that year he began work as curator, librarian and indexer at the Society's apartments in Somerset House. So Mary's visit to London would have been after this date.

Mary's diary continues: 'went also to the zoological rooms, saw a quantity of stuffed animals one little creature had a maily [armoured?] coate which it had the power to lift under the belly hairy about the Size of a rat

came from Antigua also the ornithocephelas [*sic*; orni-
thorhynchus – duck-billed platypus] with its spoon bill
[sketch of ornithorhynchus] also the skeleton of a bat
some of the bones much resembling my pterydactyle.'
She added: 'thought Regent Street and quadrant fine, also
Somerset place beautiful.' The Regent Street development, with Londoner John Nash in overall charge, had
been completed about 1820, and was a sight not to be
missed at the time of Mary's visit. The sweeping curve of
Regent Street between Piccadilly and Oxford Circus,
known as The Quadrant, was intended for fashionable
shops. The elegant Quadrant with its colonnades of
columns met with Mary's approval. Nash had also
designed the reconstruction of Buckingham Palace in
1825, and was about to start on Trafalgar Square.

Mary also visited the British Museum, home of the
Natural History collection at that time, about which her
departed young friend, Miss Bell, had written to her so
enthusiastically back in 1824. The museum was also in a
period of transition. New buildings had been started in
1823–6 for the King's Library. Next came the west wing in
1831–4, followed by years of work on the rest of
the structure.

On Thursday went to the [British] Museum with wich
[*sic*] I was much delighted beside the old Library
[British Library] I saw the King's Library [donated by
George III in 1823] in which are sixty thousand
Volumes, saw a Book of Queen Elizebeth's [*sic*] writing
also King Ethelbert's prayer Book Lady Jane Greys do
[ditto], also an illuminated missel [missal] of Richards
the Second. Henry the 7th prayer Book Queen Mary's
the 1st. do, all in a high state of preservation . . .

'Curious and beautiful Beyond description' was her assessment of the King's prayer book.

> Saw the Elgin Marbles thought them remarkably fine, the Gigantice [sic] egyptian heads of which the expression was more than fine mummies coffins do & hethan [heathen] Gods without number; the two pillars of Granite in the K Liby [King's Library] cost two thousand £ each. Saturday: Baker Street Bazar [sic] in wich [which] was the Diorama of St Peters at Rome the panorama [following written by someone else] of Rome, Bohemian brothers [from Czechoslovakia] sang, and to Mr Sowerby's museum.

It is not surprising that Mary went to Sowerby's. George Brettingham Sowerby had, of course, been her agent since 1824. He and his son were conchologists, who sold a variety of natural history specimens, including fossils and minerals. Mrs Murchison also kindly looked out for Mary's best interests, helping her to make contacts and sell fossils. In a letter to her friend Mrs Murchison earlier in the same year, Mary wrote, 'I am also greatly obliged to your kindness in mentioning my skeletons to different Collectors.' Sedgwick and Lyell were two of those Mrs Murchison contacted on behalf of Mary. Charles Konig, the Keeper of Natural History at the British Museum, was another. In February 1829 he requested a sketch of a 'round vegetable' from Mary, and she referred him to the Engravers for the Geological Society.[28] Sowerby's were situated on King Street, leading in to Covent Garden Market, with all its colours, smells and sounds. It, too, was changing; from 1828 to 1831 a new market building was constructed, after which the mix of characters and the liveliness of the place attracted fashionable society.

The diary concludes, 'Went to St Mary's church Heard Dr Dibdin preach.' The rector, Dr Dibdin, was Thomas Frognal Dibdin (1776–1847), from a famous theatrical family, who took holy orders in 1804. This is all that remains of her account. It deals with invertebrate animals as seen by a scientific mind, her wonder on encountering some of the sights in the world's greatest city for the first time. Then there is a more human side – being impressed by the cost of things, going shopping, hearing 'the Brothers' sing, and going to church. Of her impressions of the spectacle of London, the people, the sparkling place where wealth and want existed side by side, we know nothing.

If she visited the Trafalgar Square area, so-named from 1835, she would have found it, too, in a state of flux. From 1830, demolition was under way to create the square. The 1805 Battle of Trafalgar was some time past, but Nelson was not yet surveying the scene from the top of the 145-foot fluted Corinthian column, nor had the granite fountains and four bronze lions arrived.

The lions, however, were already there. Mary would have been fascinated to know that 3–4 yards below the modern ground surface, in a hollow in the Thames gravel, were the fossils of prehistoric mammals: lions, hippopotami and other beasts, from a straight-tusked elephant to a narrow-nosed rhinoceros, whose remote descendants are now to be found in zoos and in East Africa. These finds were discovered during excavation work in the 1950s. It would be a similar story in many other areas of London, where deep digging would reveal fossils from prehistoric marine life to rhinoceroses and woolly mammoths. An elephant-like mammoth, relative of the mastodon, had famously been found at Charing Cross in 1690. And this recalls novelist Louis-Ferdinand Céline's remark, in the 1930s, that the bunched-up buses in Piccadilly Circus reminded him of a 'herd of mastodons' reinhabiting their prehistoric landscape.

Perhaps Mary Anning walked near St Paul's Cathedral as the lung-choking smog from a million coal fires swirled, imagining a scene similar to Charles Dickens's opening in *Bleak House* (1851): 'As much mud in the streets, as if the waters had but newly retired from the face of the earth, and it would not be wonderful to meet a Megalosaurus, forty feet long or so, waddling like an elephantine lizard up Holborn Hill.' Dickens liked to keep up to date with scientific advancement. He knew about the *Megalosaurus*, a large and powerful predator with teeth serrated like steak knives. In 1824, it had been the first dinosaur to be named, as noted, by Buckland and Conybeare. The length was originally thought to have been much greater than it proved to be, although it was still the size of a London double-decker bus.

Charles Dickens also knew of Mary Anning. In 1865, almost twenty years after her death, he included an article about her in *All the Year Round*. In it he praised her 'good stubborn English perseverance', her intuition, her courage, physical and mental, in the face of those locals who initially mocked her eccentricity. The specimens she found enabled the wise men to identify 'four kinds of icthyosaur . . . two plesiosaur . . . and the extraordinary pterodactyle . . . which made Cuvier . . . award the palm of strangeness to a monster half vampire, half woodcock, with crocodile's teeth along its tapering bill, and scale armour over its lizard-shaped body', as well as talons at the anterior joints of its wings. She did much more besides. Dickens concluded his overview of her life with, 'the carpenter's daughter has won a name for herself, and has deserved to win it'.[29]

*   *   *

Back home in Dorset, even in little tucked away Lyme Regis, changes were sweeping in. One of the most important national developments during Mary Anning's lifetime was the passing of the first Reform Bill in 1832 and the turbulent events leading up to it. After the war with the French ended in 1815, the English – including geologists – were once more able to travel to France. But the war had cost Britain over £1 million, with little to show for it. Times were hard. Now problems at home came to the fore.

One effect of the French Revolution on life in England was the negative impact it had on the seedling issue of women's rights. People were more determined than ever to cling to the old conventions. Mary Wollstonecraft had written *A Vindication of the Rights of Women* in 1792, but progress did not begin until after Mary Anning's death, in the second half of the nineteenth century.

The years 1814–15 were marked by a severe agricultural depression. Kind-hearted Buckland, on his travels around the country, was aware of the agricultural distress and looked for ways to help. He took action by buying several farms on which he experimented with better drainage, crop rotation and so on, and worked to improve the soil and agricultural conditions. Mantell, too, although, on the one hand, he tried to ingratiate himself with the nobility – for his own survival – on the other, always steadfastly defended those who had no voice.

Lyme was one of the infamous 'pocket' or 'rotten' boroughs in the country. The representatives for Parliament were selected by the powerful Tory Fane family of Bristol, who had dominated local politics from 1722, while taking no interest in the town. In 1812, nineteen voters, eleven of whom did not even live in Lyme, returned two members to Parliament – while Manchester, Birmingham and Leeds had no representatives at all. In fact, the Fanes 'owned' the right to represent the borough; the 'cost' of buying a rotten borough varied considerably. A well-known example was that of William Wilberforce in Yorkshire in 1807; he had paid the enormous sum of £28,000 because of the powerful forces ranged against him and his fight for the abolition of slavery.

A popular eighteenth-century rhyme proclaimed the fame of Lyme Regis as a rotten borough:

> Satan resolved to take a rout
> And search the country round about
> To find where he could fix his seat,
> Where fraud, hypocrisy, deceit
> And avarice did mostly dwell
> To furnish candidates for Hell.
> One of his agents by his side
> With a malicious grin replied,
> 'Give 'o-er your search, 'tis wasting time –
> You'll find all you can wish in Lyme.'

The First Reform Bill had been rejected in 1831, and after the second was passed, the following election was an important occasion. In Lyme on voting day in 1832 the hustings were erected near the Assembly Rooms, and those eligible voted in nearby Jefferd's Baths. Anna Maria

Pinney's brother, 'the young gentleman', William Pinney, was the man elected to fill Lyme and Charmouth's now sole seat (79 votes, 104 votes going to the other two candidates). Pinney would be the representative for Lyme and Charmouth almost continuously for years to come.[30]

When the Reform Act became law, it enabled an additional 500,000 men in the country to vote, and redistributed parliamentary seats in a more fair and logical way, although, of course, it did not yet include women or the poor. In 1816, only thirty-one people had the vote in Lyme.[31] With the passing of the Act in 1832, this was increased to an unspectacular 183 out of a population of 3,345 (1831 census). Much more than this was wanted. Full suffrage for men and some women came in 1918, but universal suffrage not until 1928.

With extremely low wages for farm labourers, more so in Dorset than anywhere else in the country, the time was ripe for trade-union activity. In 1834 six agricultural workers were arrested for forming a trade-union lodge, requiring them to take an oath, which was unlawful. They became known as the Tolpuddle martyrs. Tolpuddle is close to Lyme, and only 9 miles east of Dorchester. Their modest aim was to fight against a *reduction* in their starvation wages. The Government feared unrest, and the men were made an example of and transported to Australia. This provoked petitions and a vociferous campaign, and they were pardoned two years later.

Pinney lost his seat in 1841 to Tom Hussey, son of the squire, and lost again in the 1847 election, when Neville Abdy won. Lyme had not left its previously questionable politics behind. There had been scandalous corruption, and a parliamentary inquiry ensued after both elections.

QUEEN VICTORIA VISITS LYME

Royal personages and events attracting large public gatherings could not be isolated from the agitation resulting from the stirrings for reform in a particularly impoverished area where year after year agricultural depression was the worst in the country. Even a tour by Princess Victoria (1819–1901) was affected, when she visited the south coast four years before she ascended the throne. Reclusive, obese George IV died in 1830, making his brother William IV king, and 11-year-old Victoria Heir Presumptive.

The fickle crowds who had cheered William on his Coronation day only one month later directed their growing fury towards the already unpopular monarch. They were agitating for reforms, necessitated by the profound changes caused by the Industrial Revolution, and expected solutions to increasingly acute social problems. The situation was further inflamed when the bill to reform Parliament was initially defeated in the House of Lords in 1831, resulting in a run on gold, and attacks on politicians' houses, jails, municipal offices and other symbols of authority.

In the years 1832–5, the Duchess of Kent arranged for the tiny future queen, aged 13 in 1832 and looking even younger, to make Royal Progresses to different parts of the country. In 1833, the party again stayed at Norris Castle, IOW. The Princess's governess, Dash her pet dog and her 'very own bed' comfortingly always travelled with her. On 8 July 1833, they made the first excursion from the castle to Southampton on the cutter *Emerald*. Accompanying the cutter was the much admired steamer *Messenger* – the first ever to be anchored at Lyme – which was on hand to tow the cutter when necessary, and also to carry the carriages and supplies (the horses were changed at staging posts). At Southampton agitation for reform had exploded only two weeks earlier, when furious boatmen demolished a new pier – the very pier the Duchess was to declare open. Superficial repairs were made, and the ceremony went ahead before a crowd of 25,000.

The next day they moved on by carriage through Dorchester to Melbury, the estate of the 3rd Earl of Ilchester. This stopover in the centre of Dorset had been carefully selected. Lord Ilchester sat in the House of Lords, and he and his family had held positions in the royal courts of both George IV and William IV. Not included in the visit were the Welds of nearby Lulworth Castle, for the first husband of Mrs Fitzherbert was Edward Weld.

Lyme Regis had sight of the young woman who would soon leave her mark on an era. Mary Anning was 34, and was both literally and figuratively breaking new ground in palaeontology for women. Victoria was 14, and would soon, as the idealised queen with her idealised husband and perfect family life, be revered as an example to all womankind.

The Princess's party drove by carriage to Lyme Regis, to meet up with the *Emerald*, on the way passing through Beaminster, Bridport and Charmouth. At Lyme the welcoming dignitaries waited in Cockmoile Square, but the royal carriage proceeded to the harbour via the Cobb Gate, following the route across the sands to the western causeway of the Cobb, where the loyal address was finally read. Mary and her mother would surely have been among the crowd. The royal party then embarked from 'the Victoria steps at the Crab-head [the harbour mouth]', where the curves of the arms of the Cobb have been likened to a crustacean's claws, and on to a naval barge that ferried them to the *Emerald*; they sailed away in the cutter, whose elegance and swiftness always thrilled the Princess. From Lyme they continued westward to Torquay and Plymouth.

One of the many boats in the harbour overloaded with cheering crowds was the *Eagle*. The crew fired a salute with their muskets, a 'feu de joie', and gave three cheers. The little Princess (as an adult she was only 4 feet 11 inches tall) was warmly received everywhere. She wrote in her leather-bound diary that she could not view the scenery very well because of the crush of excited people wherever she looked. They no doubt sensed a fresh start and an end to the extravagance and moral indiscipline of the preceding Hanoverians. Lyme, with its distinctive setting, at least got a mention in the Princess's diary: 'a small port, but a very pretty one'.

*   *   *

Smuggling along the south coast, especially in Dorset and Devon between 1772 and 1880, was rampant and almost a 'legitimate' trade, because of the wars with the French and high taxation on imported goods (tea, tobacco and spirits especially). A famous smuggler and character, bold John Rathenbury of Beer, had married a Lyme girl, and later wrote a best-selling biography about his illegal life. The *Eagle* that gave Victoria a send-off was a Revenue cutter, normally used for pursuing smugglers. Mary Anning, on the beach for so much of the day, often came upon smuggled goods that had been washed up – they were cut loose when capture was imminent – or concealed along the shore. She re-hid them so the Revenue men could not locate them, and then alerted needy locals as to where they could be 'found'.[32]

In Lyme, Mary Anning remained a pre-eminent figure. In 1839, a visitor noted:

A recent and powerful cause of attraction to visit Lyme Regis has arisen in the rich source of fossil remains . . . in the blue lias cliffs and beds in the neighbourhood . . . the principal collector of these fossil treasures now in Lyme, is Miss Mary Anning, whose Museum [shop] is one of the chief attractions of the place. Her museum contains a large collection of fossil treasures of the spot, with which it seems constantly filled . . .

And Mary was as always occupied with her business: 'every day something is sent off to some public or private collection.'[33]

Mary was well versed in the various means of transporting her finds to buyers, sensibly making use of whatever method was available. She sent larger specimens by ship and later by railway. Although the great boom in railway building was from 1830 to 1840, the railway did not reach Axminster until 1859, more than ten years after her death.[34] In what turned out to be Mary's penultimate letter to Sedgwick on 20 May 1843, she stated that she was sending off the 'platyodon head' – probably the extremely rare *Temnodontosaurus risor* (juvenile *T. platyodon*) – by 'waggon railroad to London'.[35] She might also send a letter with someone in Lyme who she knew would be passing near to the intended recipient.

Life was moving on. There had been births and deaths, expectations and disappointments, she had visited London, seen royalty, survived devastating storms and breathed deeply on fine days after a successful search. The modest maiden who had released from the rock previously unknown giants that once roamed the earth was already becoming a legend. Perhaps there were still days on which she allowed herself to dream, but she knew that the pattern of her life was now set in stone as firmly as that of her first ichthyosaur had been. But it, of course, had been set free. There were more adventures to come, as Mary's skills became ever more refined, and her luck held as she continued successfully to dodge danger in her daily life.

5

Her Spheres of Excellence

She is 'a history and a mystery'.

(Molly Anning[1])

Mary Anning's life was almost bookended by the Darwins, Erasmus (1731–1802) and his grandson Charles (1809–82). Mary was only 3 years old when Erasmus Darwin died. Charles Darwin was born when Mary was 10 – just before her discovery of the ichthyosaur. Finally, Mary herself departed this life in 1847, only twelve years before Darwin published *The Origin of Species by Means of Natural Selection*. Much of his work had already been completed when she was alive.

Although Mary made her discoveries years before *The Origin of Species* was published in 1859, she was one of those who helped prepare the way for Darwin. He could have published earlier, in Mary's lifetime – he had completed his initial preparation as early as 1842. However, the delay, he later acknowledged, gave people time to allow their minds to adjust to the new thinking. The enormous amount of data he had collected would demonstrate that species were not fixed but evolved by a process of natural selection. And Mary Anning, with her network of professional contacts and her enquiring mind, must have known the direction in which the new thinking was moving.

Charles Darwin's polymath grandfather Erasmus is very much part of his story. This extraordinary character had anticipated the views of Jean Baptiste de Lamarck (1744–1829) (the leading French expert before Cuvier) – and those of grandson Charles. Erasmus – inventor, physician, scientist and poet – wrote about evolution in his long poem *The Botanic Garden* (1789), and *Zoomania, or the Laws of Organic Life* (1794–6), a rhyming treatise. (Darwin's other grandfather was Josiah Wedgwood.)

The timing of Mary's discoveries fitted in neatly with Darwin's great opus. She was quietly in place, ready to come to the notice of scholars and the public for what was to be her distinctive role in the development of English geology.

*   *   *

Success in hunting for fossils was unpredictable at the best of times and there was no guarantee of finding anything of value. In the autumn of 1833 Mary wrote to Mrs Murchison, 'at present I have not anything fine or rare, for the last year I have been very unsuccessful', but she hoped that the stormy winter ahead would reveal new treasures on the foreshore. It is also possible that the easier-to-find fossils – difficult as they may have been – had all been discovered by the 1830s. Furthermore, the

economic situation was made much more difficult for
Mary when sixty banks in London and Bristol crashed in
1825–6. Then there was the repressive effect of the Corn
Law of 1825 (abolished in 1846).

Mary, who had once had regular visits from Buckland
and other scholars, now wrote, 'not having seen Dr B.
[Buckland] or any other professor for the last year I know
no more of what is going on in Geology than the man in
the moon'.[2] And in 1844 she wrote, 'Lyme being such an
out of the way Town, I seldom get any scientific news'.[3]
But she always kept in touch with her circle of remarkable men and women. Perhaps it had just been a bad year
for professors' visits. In 1892, a lady from Lyme recalled:
'Fifty years ago! Lyme was then as much a winter resort
as a place to pass a summer holiday and the round
gaities, festivities, and the constant presence in town of
men of mark were immunities from stagnation. On a fine
day in mid-winter the streets presented as animated a
scene as they do now in mid-August. The eminent geologists, Buckland, Conybeare, Sir H.T. De La Beche and
others in the scientific pursuits were frequent visitors.'[4]

Mary's past success was due to a number of reasons.
She was persistent, and physically she was tough. She
could not have accomplished all that she did otherwise.
And her determination to keep going was part of it.
Pinney wrote: 'she has supported her mother and brother
in bitter poverty and [even] when she was so ill she was
brought fainting from the beach.'[5] Years of experience
had given her a highly developed instinct for looking in
the right place after a storm: 'She has more than the
power of an eagle's eye when she searches among the
sands and rocks, and can distinguish when fossils are
enclosed in the stone.'[6] George Roberts summed up Mary

Anning's wide-ranging skills by saying she had 'a genius for discovering where the Ichthyosauri lie embedded', as well as 'great judgement in extracting the animals, and infinite skill and manipulation in their development'.[7]

Just as important, and even more unusual, were her skills in dissection and preservation. She dismembered existing creatures to compare them to similar fossilised animals using comparative anatomy. One, a marine gastropod – the sea hare, *Aplysia punctata* – she compared with fossilised shells of belemnites. From this she was able to deduce the association of the belemnite and its ink sac, later proven when complete fossils of the sea hare were unearthed.

When she found the strange fossil fish, *Squaloraja*, the experts thought it was a ray. But Mary, armed with her experience, knowledge and skill at dissection, dismembered a modern ray to discover they were different: she had found another new species. Similarly, she looked for features that both pterosaurs and bats had in common.

Often noted were the great care and time Mary took in preserving and presenting the fossils, so they remained as complete as possible for the scholars to study. How was it possible that this uneducated person, a woman, could possess the knowledge she seemed to display? It was difficult for educated geologists and others to understand – especially when she was younger. One visitor to Lyme in 1824 was Lady Silvester, the widow of a former Recorder in the City of London, an eminent position befitting the man who was the senior law officer in the City and the senior judge at the Central Criminal Court. She wrote in her diary after visiting 'the famous fossilist reconed [*sic*] the best in England':

The extraordinary thing in this young woman is that she has made herself so thoroughly acquainted with the science that the moment she finds any bones she knows to what tribe they belong. She fixes the bones on a frame with cement and then makes drawings of them and has them engraved . . . It is certainly a wonderful instance of divine favour – that this poor, ignorant girl should be so blessed, for by reading and application she has arrived to that degree of knowledge as to be in the habit of writing and talking with professors and other clever men on the subject, and they all acknowledge that she understands more of the science than anyone else in this kingdom.[8]

Lady Silvester's solution to the puzzle of Mary Anning's intelligence was to invoke 'divine favour'.

Mary's skill at drawing the specimens she found has also been noted. Some of the drawings remain, among them her sketch of the *Ichthyosaurus*, and the drawing of the *Squaloraja* that she sent to Professor Adam Sedgwick at Cambridge University, when she was attempting to interest him in purchasing it.

Professor Sedgwick was one among many eminent collectors and institutions to whom Mary sold fossils. By 1841 he had persuaded Cambridge University to establish a formal museum in one of the existing buildings; after his death his friends and colleagues, as their memorial to him, opened a public subscription to raise the funds for the present museum. From the beginning the specimens accumulated faster than the space available at Cambridge. Sedgwick had built up one of the finest geological research museums in the world, which now holds 1.5 million fossils, rocks and mineral specimens.

In the archives are the eight letters Mary wrote to Sedgwick, discovered by Hugh S. Torrens. These were added to by three more, found by history graduate Philip Dunn in the 1980s; there are also Sedgwick's 'rough' account book in which he recorded transactions and the Woodwardian accounts with confirming entries and vouchers. The record is still incomplete. As usual, benefactors were carefully credited, to encourage future donations – but not the person who actually found the specimen, especially one that was purchased. That was simply a business transaction.

Although her letters mainly concerned such transactions, they add valuable insight into her work. She discussed either what she had already sold to Sedgwick, or what she was offering for sale. This was an accomplished saleswoman, who used winning phrases to promote her finds to a possible buyer: 'the best yet discovered' and 'worthy of a place in a Museum', for example.[9]

*   *   *

Contradictory descriptions of Mary's personality remain. Among the male scholars, opinions varied, but the most consistent was that they respected her knowledge and regarded her as an equal in geology, even if some were reluctant to give credit where it was so obviously due.

It is good to know that she enjoyed an argument, did not suffer fools gladly and could hold her own. 'I do so enjoy an opposition among the big-wigs,' she wrote to Charlotte Murchison in February 1829.[10] Yet Mary wanted to be a person of consequence. Her 'proud and unyielding spirit', as Pinney described it, must have been a great strength; to some it was an unwelcome feature, but it no doubt helped to ensure her survival. There was an element of Mary's personality that some interpreted as self-confidence, others almost as arrogance. Perhaps she was not arrogant enough, for others noted that, at times, she was too humble. John Murray accompanied her on a foreshore walk, on which she wore 'an appropriate, though sufficiently curious' outfit, while he was the 'Alpine Jäger'. He carefully watched his step so as not to fall, as he had heard that shortsighted Lord Cole had done so several times on such walks with Mary. John Murray later wrote that at the end of the day, she handed over to him all the finds they had collected, at no charge, and he commented: 'Thus curious are the simple annals of the poor.'[11] At times she wished to curry favour, at other times to assert herself. Or perhaps she was able to release some of her frustration in this way by taking a hard, assertive line when recounting events to Miss Pinney, which she was in fact seldom able to do in practice.

Always, she was too open-handed in sharing knowledge. She worked as a cliff and foreshore guide, showing visitors and professional men the best places for collecting – as indeed did the Philpots. A contributor to *Chambers's Journal* wrote: 'Miss Anning . . . braved all weathers, and was far too generous in allowing even wealthy visitors to accompany her in her explorations without charging a fee, as some naturalists now reasonably do.'[12]

In 1832 Miss Pinney recorded that Mary spoke her mind to whomever she pleased: 'She has been noticed by all the cleverest men in England who have her to stay at their houses, correspond with her on Geology &c. This has completely turned her head and she has the proudest and most unyielding spirit I have ever met with. Much "learning has made her mad". She glories in being afraid of no one, and in saying everything she pleases. She would offend all the world, were she not considered a privileged person.'[13]

Privileged (to a point) she may have been, in Pinney's eyes, but bitterness was creeping in. Blunt speaking and wilfulness may also have been the prerogative of someone with an inward anger that brought on black moods, and to Mantell she was 'vinegary'. She may have felt aged beyond her years or at the very least an old maid, knowing that nothing would change now.

Charles Dickens suggested, sometime after her death, that she had *not* been a 'prophetess' to the locals; instead she had been 'derided and laughed at when she began her researches', and later had been 'laughed at as an uneducated, assuming person who had made one good chance hit'.[14] Even today, this attitude would not be surprising in an isolated small town, when applied to a local impoverished spinster, perhaps somewhat masculine looking, oddly dressed, who walked the shoreline in all weathers, sold the strange fossils she found, and yet was the centre of attention.

She could be kindly and gentle, but there was another side to her personality. She relished gossip, was fearlessly disparaging about many of the local people, and her sharp tongue could deliver a memorably stinging rebuke.

George Roberts hinted at an unpleasant feature of her personality. In his summing-up of her life in *History and Antiquities of the Borough of Lyme Regis* (1834), Roberts intimated that there was something he did not want to say, while admitting that she would not mind if he did say it: 'Though the subject of this notice would not be disparaged by a description more strictly personal, yet such might be unpleasant.'[15] What could it have been? She was only 35, so it was not the crankiness and oddities of old age. Perhaps it was her prickliness or her appearance. Anyone might have been irritable having lived an unremittingly hard life, and knowing that others had received most of the credit and benefit from her efforts. Clearly there were facets of her character that she suppressed or kept hidden when necessary, which, perhaps, in a way made her a more normal rounded person. Miss Pinney noted on 25 October 1831: 'she says the world has used her ill and she does not care for it, according to her these men of learning have sucked her brains, and made a great deal by publishing works, of which she furnished the contents, while she derived none of the advantages.'[16]

Her loyalty to her friends was paramount, and she exhibits a refreshing sense of humour in a poem she wrote at a time when she was becoming seriously ill. In 1846, her old friend Roderick Impey Murchison was knighted by the Prime Minister, Sir Robert Peel. Mary wrote a congratulatory poem to him, reproduced here in full because it demonstrates her perceptive humour and loyalty to a friend. She prefaced it with: 'I am glad the country has confirmed the honours of Russia [where the Emperor had given him two Russian orders when, after lengthy study, he published his book on the geology of Russia and the Ural Mountains] . . . He won them well, and may he wear them long.' The poem is entitled 'Encomium Murchisonarum first'.

Who first survey'd the Russian states?
And made the great Azoic date?
And work'd the Scandinavian slates?
      Sir Roderick.

Who calculated nature's shocks?
And proved the low Silurian rocks
Detritus of more ancient blocks?
      Sir Roderick.

Who knows of what all rocks consist?
And sees his way where all is mist
About the metamorphic schist?
      Sir Roderick.

Who draws distinctions clean and nice
Between the old and newer Gneiss?
And talks no nonsense about ice?
      Sir Roderick.

Let Sedgwick say how things began,
Defend the old Creation plan,
And smash the new ones if he can,
      Sir Roderick.

Let Buckland set the land to rights,
Find meat in peas, and starch in blights,
And future food in coprolites,
      Sir Roderick.

Let Agassiz appreciate tails,
And like the Virgin hold the scales,
And Owen draw the teeth of whales,
      Sir Roderick.

Take then thy orders hard to spell,
And titles more than man can spell.
I wish all such were earn't so well,
                    Sir Roderick.[17]

The humorous mocking of the other fine gentlemen –
including Sedgwick defending the 'old Creation plan' –
reveals a much lighter side to Mary. She was fortunate in
having friends like the Murchisons, who were kind to her
in many ways and obviously had a sense of fun. Charlotte,
Lady Murchison, noted for her charm and social skills,
was a distinguished geologist in her own right. She went
on fieldwork trips with her husband, playing a full role in
fossil-hunting. She had been well taught in the practical-
ities of collecting and sketching by Mary Anning, as Sir
Roderick later said after his wife's death.

*   *   *

Although she received little formal education, Mary was
far from being uneducated. There was, of course, her
superb practical education on the foreshore by the cliffs
every day. Also, she had a few books of her own, and
copied out scholarly articles by Conybeare and others, the
illustrations competently reproduced as well, to keep by
her for reference. She was even obliged to copy out
Conybeare's article announcing her own find: 'On the
Discovery of an Almost Perfect Skeleton of the Plesiosaurus' in 1824. As she was doing so, faithful Tray, lying
near her, caught her eye, and she made a sketch of her
beloved companion looking up at her. In Lyme, a library
and two bookshops – aimed at the visitors – were close
by. There is also a suggestion that she tried to teach
herself French in order to read Cuvier's works.[18]

Mary Anning also kept notebooks, which she called her Commonplace Books; sadly, only 'No. 4' has survived and is now in the Dorset County Museum, Dorchester. It reveals the range of her interests, and the thoughts and prayers that were so special to her they were worth the effort of such a busy person writing them down. She had, close by, prayers suitable for all occasions – for Sundays, for a sick room, for a lingering illness; Pinney noted 'she is gentle, attentive and has a simple . . . way of expressing her affectionate feelings, beyond any person I ever met with . . . she will attend the sick both night and day, when they are ill with infectious diseases'.[19]

Byron had died in 1824. A copy of the last poem he wrote, 'On This Day I Complete My Thirty-Sixth Year', was also among her papers. It begins,

> 'Tis time this heart should be unmoved,
> Since others it has ceased to move;
> Yet though I cannot be beloved,
> Still let me love!

The Murchisons had toured the Continent for over two years (1816–18) following their marriage, to save money! In Switzerland, on one of their vigorous walks, they had bumped into Byron; that very night there was a fierce storm after which Byron wrote *Childe Harold's Pilgrimage*. Mary no doubt heard an account of this chance meeting. And from 'The Magdalene' came:

> Oh turn not such a withering look
> On one who still can feel,
> Nor by a cold and harsh rebuke,
> An outcast's mis'ery deal!

There were verses from hymns, one verse of Gray's 'Elegy Written in a Country Churchyard' (1751), and moral maxims from the 'Swan of Lichfield', Mrs Anna Seward, in her 'Letters', for example: 'Nothing but an independent fortune can enable an amiable female to look down without misery, upon the censures of the many, and even in that situation their arrows have power to wound, if not to destroy peace . . .'. Mary collected prayers for morning and evening as well. On loose sheets she had copied out an article on the planets, various Victorian verses, a poem entitled 'The Virgin' by James Montgomery (1771–1850), another, 'To a Bride', which may have been by Mary, and an essay entitled 'Woman!', which was indeed by Mary:

> And what is a woman? Was she not made of the same flesh and blood as lordly Man? Yes, and was destined doubtless, to become his friend, his helpmate on his pilgrimage but surely not his slave, for is not reason Hers? Are not her claims 'To share redeeming love' as great? . . . Woman seems throughout the sacred scripture . . . *more* than even man the object of this pure benevolence. And woman (when his own disciples fled and left him) dared to attend his cross; they were his constant followers.
>
> And women too were honor'd with the message given by the bright ambassadors of heaven . . . Say then shall woman sink beneath the scorn of haughty man? No let her claim, the hand of fellowship . . .[20]

Perhaps Mary's steadfast belief in God gave her the courage to carry on, to continue to explore the dangerous cliffs. Her friend Anna Maria Pinney, who earlier had noted her piety, wrote in December 1833 that Mary had had several close escapes from danger: 'The word of God is becoming precious to her after her late accident, being nearly crushed to death. I found it healing her mind.'[21]

Mary and her mother, Molly, were fond of each other, and lived together for all her mother's life. There is a description of 'Mrs Anning the Fossilwoman's mother', when a very old lady, wearing a mop cap and large white apron. Children sometimes saw Molly enter the shop with feeble steps to help them choose a little treasure. It was a dim recollection for the writer many years later, who made a point of noting that 'the two were devoted to each other'.[22] To mother Molly, her daughter was 'a history and a mystery', summing up a complex personality as perhaps only a mother can. Molly died in early October 1842. At the age of 43, Mary was now on her own. A lonely life must have become even lonelier. In 'No. 4' Notebook are several verses by Henry Kirke White, one entitled 'Solitude', copied out after her mother's death:

> It is that I am all alone . . .
> Yet in my dreams a form I view
> That thinks on me and loves me, too;
> I start and when the vision's flown,
> I weep that I am all alone.[23]

Other poems are consoling: her 'Almighty Friend' knows her fears and her tears.

The role of women in a scholarly environment came to the fore when the British Association for the Advancement of Science was formed in 1831 with Buckland as president. A main point of discussion was whether women should be admitted. The eventual answer was 'no', Buckland explaining: 'Everyone agrees that, if the meeting is to be of scientific utility, ladies ought not to attend the reading of the papers and especially at Oxford as it would at once turn the thing into a sort of Albemarle-dilettante-meeting, instead of a serious philosophical union of working men.'[24] And this was from a man with an extremely intelligent wife.

*   *   *

Those searching for a spinster's secret sex life can ponder the gossip hinted at by Cumberland in 1824 when Colonel Birch sold his collection to bail out the Annings. The next contender was the good-looking William Buckland. His daughter, Elizabeth Gordon, recorded that people wondered what the relationship had been between Mary and her father. She wrote in her biography of him: 'The vacations of his earlier Oxford time were spent near Lyme Regis. For years afterwards local gossip preserved traditions of his adventures with that geological celebrity, Mary Anning, in whose company he was to be seen wading up to his knees in search of fossils in the blue lias . . .'.[25] In 1844, after Buckland's youngest child, Adam, died, he moved his family to Lyme.[26] Mary had long been a friend to the children.

As for what was appealing to her in masculine looks, Mary knew what she did like, and what she did not. Of the popinjays who visited the seaside at Lyme, she said that they were 'things or numskulls, not men'.[27] Her idealised man seems to have been Sir Roderick Murchison. In a letter to Mrs Murchison, in October 1833, Mary wrote that she had not yet received a copy of

Mr Murchison's last anniversary speech; she had heard that he 'looked like a God' when delivering it, 'which I most cordially believe for [he] is certainly the handsomest piece of flesh and blood I ever saw'.[28]

Mary does seem to have been overwrought at times. There was a strong hint in Pinney's diary entry for 8 November 1831 that something dramatic may have happened to her: 'Been out yesterday and today with Mary Anning fossilizing. Miss Lucy Oates went with us today . . . As soon as she left us Mary Anning turned round to me and said, "That young lady has been crossed in love." This woman must have penetration to have discovered such a thing without any conversation except on Geology.'[29] This seems to be an indication of the emotional state of Mary herself. Something had apparently gone wrong, especially in light of the following, written by Pinney on 23 January 1832:

Her [Mary Anning's] wonderful history (which I cannot consider myself at liberty even to write) interested me . . . I felt the power of the emotions by which she was actuated, and I should have been glad to have possessed sufficient strength of mind to have done the same. An illness of eight years could not bend that spirit, though acute pain supplied the place of health, the bodily anguish was small with what must have been suffered by a proud mind, who had hoped since childhood to see herself removed from her low situation in life, and suddenly saw those hopes blasted by Satanic treachery.

These were strong words from a gentle and self-contained woman. At the same time, Mary told Anna Maria that a

Mrs Oates had asked her mother whether Mary had ever been engaged to a man; 'old Mrs Anning' replied that 'her Mary had no sweethearts, and she did not suppose that she went talking about that there nonsense to Miss Pinney'.[30]

What could have affected Mary eight years earlier? What *was* the 'Satanic treachery'? Henry De La Beche, a friend since childhood, is a possible runner. Was it a coincidence that he obtained a legal separation from his wife in 1825; or that, in the same year, William Buckland finally married, aged 40? Both events happened eight years before the entry in the journal in which Anna Maria mentions Mary's 'illness' of eight years. Speculate as one may, there is no proof for any of these theories. Then, on 19 February 1832, Pinney continued: 'Heard from Mary Anning – poor thing – she feels the misery of her hardened heart and almost fears the door of mercy shut.'[31] It seems clear that there was some attachment in which she had been badly let down.

Plain in looks and with a strong individualistic character, Mary Anning was 33 years old at a time when most women aimed to be married in their late teens or early twenties. She knew there would be no further chances to escape from the unremitting labour in the cold winter months, and the near poverty, or to enjoy a higher social status, to have her work properly acknowledged, or to live the life of ease enjoyed by her privileged visitors. However, there is also some suggestion – and it is only that – in Anna Maria's journal that Mary did not get on with her sister-in-law, Joseph's wife. Perhaps that was the cause of the upset.

*   *   *

George Roberts was perhaps the first to outline Mary Anning's accomplishments in some detail. Lyme's historian and schoolmaster knew Mary well, and described her as a 'living worthy' in his 1834 book *Roberts's History of Lyme Regis and Charmouth*. And she herself subscribed to his publication.

Although Mary had a hard life, she was also lucky. She had managed to avoid danger on a number of occasions over the decades. Owen mentioned a narrow escape when out fossiling with her, one of several near disastrous encounters with the stealthy sea. Fossil collecting was most rewarding, but particularly dangerous, in the winter months, when new specimens were usually exposed because of the turbulence of the weather and the ocean. Physically strong and with an independent spirit, Mary knew when to respect the awesome power of nature, not least the incoming tides and sudden rockfalls, as she walked and waded under unstable cliffs at low tide, sharp eyes seeking out specimens dislodged from the rocks by the latest tide. The next tide might well forever destroy, cover up or sweep away any find. Confined on one side by steep cliffs, on the other by sea, and intent on her work, she was nearly trapped several times by the returning rush of water that was as cold as death. Indeed, the tides along this coast had swept away a number of people over the years.

Edward Pidgeon in *The Fossil Remains of the Animal Kingdom* writes of the 'arduous and zealous exertions of this female fossilist in her laborious and sometimes dangerous pursuit'.[32] The dangers were acknowledged in a footnote to an article by George Cumberland, in which Mary Anning was identified as the finder of the 1821 *Ichthyosaurus*:

ROBERTS AND LYME

George Roberts (1804–60) was Lyme Regis's first schoolmaster, historian, mayor from 1848 to 1854, and a contemporary of Mary Anning. He wrote a number of books, among them several notable ones on Lyme. He was ever keen and imaginative in promoting Lyme. He noted that 'the fossilist and geologist look upon this place as the sportsman does upon Melton Mowbray [fox hunting]'. (Coincidentally, Melton Mowbray was the place where Murchison had ensconced himself to ride to hounds, before his wife redirected his enthusiasm towards geology and fossils.)

To emphasise the purity of the water in Lyme, Roberts claimed that it would be difficult to find a frog in the parish because there was no stagnant water. He promoted the merchants, stating that the shops were well stocked, and 'our milliners do not . . . introduce the fashions trois mois apres l'explosion'. Furthermore, gentlemen who have been abroad 'say Lyme has the appearance at a distance of a Turkish town', a tribute to the trees and gardens, and the 'irregularity of the houses'. Roberts blended Turkey into Greece, continuing that 'on entering Lyme one can meet such groups of Grecian beauties who have not been captives but captors'.*

Overblown Victorian language being what it was, George Roberts's only intention was to record what he knew about Lyme, all that he could find out, and to do the best for the town, and to him we are indebted for the first history. Surprisingly, he allowed himself to be made a Freeman of Lyme by the Fanes in 1831, along with fifteen others, so he could vote for their candidate in the 1832 election. In the event, such recently created Freemen were disqualified and the Fane candidate lost. No one more fully deserved to become a Freeman of Lyme than Roberts, and it was unfortunate that it occurred under such blemished circumstances.

---

* George Roberts, *Roberts's History of Lyme Regis and Charmouth*, London, Samuel Bagster and William Pickering, 1834, pp. 179–80.

This persevering female has for years gone daily in search of fossil remains of importance at every tide, for many miles under the hanging cliffs at Lyme, whose fallen masses are her immediate object, as they alone contain these valuable relics of a former world, which must be snatched at the moment of their fall, at the continual risk of being crushed by the half suspended fragments they leave behind, or be left to be destroyed by the returning tide:– to her exertions we owe nearly all the fine specimens of Ichthysauri of the great collections . . .[33]

In a letter to Charlotte Murchison in February 1829, Mary described one particular incident. Acknowledging her 'old passion for bones', she described how one day she was on the foreshore, concentrating on retrieving part of a plesiosaur. The man whom she employed to help her failed to alert her that the tide was flowing in with threatening speed. Fortunately Mary's luck held and they managed to escape. When they reached her cottage looking like 'a couple of drowned rats', she asked him why he had not warned her. He was ashamed to admit that he, a man, was frightened when she, a woman, did not appear to be! In a footnote to the letter, Mary apologised for the way it was written, adding, 'I beg to say that I have walked ten miles today and am so tired I can scarcely hold the pen.'[34]

Rockfalls along this part of the coast were not unusual. Mary's faithful companion on her lonely days out fossiling was her small dog, Tray, a black and white terrier-type of mixed breed. 'After every gale or heavy sea disturbance, the masculine figure (but always smiling face) of Mary Anning was seen with hammer and rush vasculum [collecting basket] followed by her little black and white terrier, making towards the lias beds in search of the cast up or bared antediluvian treasures . . .'[35] When she found something of interest that was too large to carry or to excavate, she left Tray to mark and guard the spot, so the story goes. Then she went to find others to help her dig the fossils out. And Tray was there beside his mistress in both the pastel and oil portraits of Mary. One day in October 1833, a rockfall very narrowly missed her, but killed loyal Tray, who was, as usual, close by. She wrote to Mrs Murchison: 'I would have answered your kind letter by return of post, if I had been able. Perhaps you will laugh when I say that the death of my old faithful dog quite upset me, the Cliff fell upon him and killed him in a moment before my eyes, and close to my feet, it was but a moment between me and the same fate.'[36]

Even the daily traffic in the town could be a threat. Tiny Cockmoile Square, where Mary grew up, was adjacent to the extremely narrow – originally only 8 feet wide – main thoroughfare of Bridge Street, dangerous to adults as well as children. In the winter of 1833 Miss Philpot noted 'one of her [Mary's] miracle escapes'. She was 'going to the beach before sunrise and was nearly killed in passing over the bridge by the wheel of a cart which threw her down and crushed her against the wall'. Fortunately, the cart stopped in time and Mary was 'not prevented from pursuing her daily employment'.[37] Bridge Street, although widened in 1913, is still very narrow, and wise pedestrians and drivers must remain alert.

Most notable of the several accounts of her narrowly escaping injury or death was the very first one, when Mary was but an infant. On this occasion, her close escape seemed to mark her out for greatness, at least with the benefit of hindsight. On 19 August 1800 she was struck by lightning, which killed the three women who were with her. There are several accounts of this incident. A crowd had gathered in the early evening, a short distance outside town in the Rack Field to watch an equestrian display. According to Jo Draper in *Mary Anning's Town – Lyme Regis*, two fields were called 'Rack Close' at that time. The most probable site for this event was a field off the Charmouth Road, where political rallies were held in the 1840s (oddly enough, close to today's Anning Road). Among the throng was a woman holding tiny Mary, accompanied by two girls about 15 years of age. A large number of people were gathered because they had heard all the talk about the previous evening's wonderful display of 'vaulting'. But threatening clouds on the intensely hot and sultry evening indicated an approaching thunderstorm.

During the performance, the rain suddenly lashed down, and some of the crowd left for home while others ran for cover under a group of elm trees. Warned of the danger of doing this, about twenty people left their leafy shelter, but not those with the infant Mary. 'A vivid discharge of the electric fluid shortly ensued, followed by the most awful clap of thunder that any present ever remembered to have heard.' A man shouted and pointed to people lying on the ground under a tree. Others ran to the spot, where the three women and a child lay motionless.[38] When 20,000 amps of electricity found the path of least resistance down an elm tree the group must have been standing within 6 feet of the tree trunk. Its deadly route could be seen burned down the bark, from which it instantaneously jumped to the right side of Mrs Elizabeth Haskings. She had been holding the toddler and must have taken the full force, which stopped her heart. The girls, Fanny Fowler and Martha Drowler, were also killed.

Why was 60-year-old Elizabeth Haskings taking care of baby Mary? Her husband, John, left a note twenty-one years after the event, explaining that the infant was a sickly child and his wife often took her out for air. He concluded, 'it may be said that the death of the Nurse was the life of the child'.[39] At first, the tiny, blackened girl seemed to have expired too.[40] She was taken to her parents apparently dead, but then placed in warm water, no doubt rubbed vigorously, and after half an hour made a full recovery. That child was Mary Anning. At just 15 months old she was already newsworthy.

Most accounts agree about what happened, except for some slight details in Miss Pinney's version, which she presumably heard from Mary herself. According to this account, Mary's mother was haymaking and gave the infant to another woman in the hayfield to hold. There was no mention of the equestrian display. When Mary Anning became a famous local character – then a national and international one – this terrifying event was recalled knowingly. It was obvious to all that she had been spared in order that she might accomplish something special. 'She had been a dull child before, but after this accident became lively and intelligent and grew up so.'[41]

Apparently there had been another close call not long after this. According to Mary, her nurse had dropped her in the sea and she had stopped breathing, but again a tepid bath had brought her round. And she also told Murray of the time when she was undermining a cliff, trying to extract an ichthyosaur, when it collapsed on her and temporarily buried her.[42]

Destruction by fire was always a possibility in the cramped, wooden-framed and mostly thatched houses of Lyme. Mary Anning's older sister, also Mary, had perished in 1798 in a house fire when only an infant. The fire engine was stationed nearby in what used to be the lock-up in Cockmoile Square. Another fire broke out in Lyme in 1803, four years after Mary's birth. It started the night before Guy Fawkes's day in a bakery in Monmouth Street. Gale-force winds whipped up the flames, destroying forty-two houses in the working-class quarter of Lyme, that is, the upper town around Mill Lane and the mill itself. The Independents' Chapel was saved from 'the devouring element', after a great effort by the locals. Once the flames had been extinguished, funds were raised for the many poor people whose homes had been destroyed, and 'a furnace' set up in the church porch for some time afterwards so they could cook their dinners.[43]

This fire was witnessed by Jane Austen, who had been lodging nearby. Five years later, on 7 October 1808, she referred back to the Lyme fire in a description of a conflagration that occurred during a visit to Southampton: 'The flames were considerable, they seemed about as near to us as those at Lyme, & to reach higher. One could not but feel uncomfortable, & I began to think of what I should do, if it came to the worst.'[44]

In 1844 there came another devastating fire, almost an echo of that of 1803. By then Mary Anning's life was moving into its final years, and this fire also marked a kind of end for the Lyme Regis she had known so well. Most of the old town was destroyed, much of the affected area being close to Cockmoile Square. By this time Mary and her mother were living at the top end of Broad Street, and her brother and his family were living nearby in

St Michael's Street, now Pound Street. The fire started on St George's Square in the historic George, the town's most important inn and once the market place. The famous old Three Cups Hotel next to the Custom House, where Tennyson had stayed, was consumed in the flames, as was the medieval Custom House itself, so central to the commerce of Lyme. The Lymites were fearful that the whole town would go up in flames. Women rushed to gather their valuables together, and then wetted blankets to fight the approaching flames, fanned by a north-east wind, while men stood on the roofs of houses near the fire to knock off the burning embers. Mary wrote that, to cut off the combustion in Broad Street, the Shambles were pulled down. This old market was a characterful 'island' of shops right outside the front door of Pryne House, No. 10 on the south side of Broad Street, the most likely address where Jane Austen had stayed back at the beginning of the century.[45] As her own life was winding down, Mary was saddened by the destruction of much that was so comfortingly familiar to her. 'I do regret the Old Clock that had stood for Centries [sic].'[46]

*   *   *

The land around Lyme was, and still is, notoriously unstable, and the shifting cliffs could spell disaster. There had been landslips before; Roberts wrote about an 'earthquake' in the area in 1689, almost certainly a landslip. In 1839 underlying springs and a great deal of rain had further destabilised an already precarious area. Mary was there at Christmas in that year when the Great Landslip occurred to the west of Lyme on the Undercliff at Downlands. Perhaps as much as 8 million tons of earth and rock slid slowly down towards the shore. This convulsion was of such size and strange beauty that the grandeur of its wild and romantic scenery still intrigues. The Axmouth to Lyme Regis Undercliff is a Natural Nature Reserve and an important wildlife area.

There had been warning signs in the preceding months, when it had been exceedingly wet. Cottagers who lived near the edges of these cliffs had been alarmed by cracks and fissures appearing in their gardens, sometimes accompanied by a sound like the snapping of tree roots. Pieces of plaster began to fall off the walls of their houses, and doors that had opened freely now scraped and stuck. On Christmas Eve a local family, the Crichards, found the path to their house sinking and alarming cracks appearing in the land round about them. Inside, the floors were buckling. They quickly removed their goods and went elsewhere, but on Christmas Day all seemed calm. Then that night a member of the coastguard had the peculiar sensation of the earth moving under him. When it happened, it was described as 'a convulsion so remarkable, from the extent, magnitude, and picturesque changes it has produced in the surface, and general configuration of a line of country, extending at least half a mile in length, by half a mile in breadth'.[47] Offshore, a reef emerged, 1 mile in length and 40 feet high, but within months it was washed away by the sea.

Among the spectators to this event were all the boys from Roberts's Crewkerne School, and with them was Frank, later Revd Francis John Rawlins, who as a boy was taught about fossils by Mary Anning. For geologists, it was fortunate that Buckland and Conybeare, still vicar at nearby Axminster, were in the area at the time of this landslide, which is also known as the Downland

Landslide or the Bindon Landslide. They visited the site immediately afterwards and each wrote a scientific account based on his own observations. Conybeare, with geologist William Dawson and Mrs Mary Buckland, produced ten plates of the event.

An account written at the time described what happened: 'It took place on Christmas Day 1839, when over forty acres of cultivated land slowly and silently slipped away to a far lower level. Two cottages were bodily removed and deposited with shattered walls to a considerable distance below the cliffs, while an orchard, which still continues to bear fruit, was transplanted as it stood.'[48]

The eye now took in the sight of broken terraces, exposed and twisted roots, crags and knolls and deep wooded rifts. Mary, too, was captivated by the raw beauty, but regretted that there were no fossils to be found because the displaced soil comprised only loose sand and chalk. In 1840 she wrote to her friend Miss M. Lister of Lincoln's Inn Fields in London: 'I have with Buckland [and] Conybear [sic] wandered all over this splendid subsidence [which they said was] worth coming five hundred miles to see . . .'. In a footnote she adds that a pyramidal crag of Culverhole Point, not more than 12 feet from the cliff, which had been a landmark, had sunk from 100 to 200 feet, and the main cliff, previously more than 50 feet from the crag, 'is now brought almost close'.[49] This disaster gave Lyme more, not less, appeal, since the thrill of a nearby landslip with its dangerous chasms and ridges fed the romantic desire for challenging, even threatening, scenery. Some visitors admired the new Undercliff from the safety of paddle steamers offshore. Reflecting the general festive atmosphere surrounding the landslide, music for a new square dance emerged: the

'Landslide Quadrille'. (The following August, thousands came to watch as a wheatfield that had been carried along with the landslip was harvested, and framed souvenirs of sheaves of wheat were sold.) Even young Queen Victoria arrived for a look.

*   *   *

Visitors of note continued to arrive in Lyme. One of these was Friedrich Ludwig Leichhardt (1813–48), a German explorer, who visited in 1837:

> stayed in Lyme Regis for 8 days . . . We had the pleasure of making the aquaintance of the Princess of Palaeontology, Miss Anning. She is a strong, energetic spinster of about 28 years of age [she was actually 38], tanned and masculine in expression. Every morning and after every stormy sea, she goes walking and clambering about on the slopes of the Lias to see whether fossils have been brought to light by falls of rocks or wave action.[50]

Born in Prussia, Leichhardt went to Australia in 1841 to explore the geology of that great continent. The 'Franklin of Australia' crossed the country from west to east, and then attempted the journey east to west, but disappeared on the tropical Cogoon River.

On a lighter note, on 1 July 1844 the King of Saxony, Frederick Augustus II (1797–1854, king from 1836), and his party visited Lyme. Dr C.G. Carus, the King's medical attendant, described a visit to Mary Anning's shop: 'We had alighted from the carriage, and were proceeding along on foot, when we fell in with a shop in which the most remarkable petrifactions and fossil remains – the head of an Ichthyosaurus, beautiful ammonites, etc. were exhibited in the window. We entered and found a little shop and adjoining chamber completely filled with fossil productions of the coast.' He continued with a description of how the fossils were found:

It is a piece of great fortune for the collectors when the heavy winter rains loosen and bring down masses of the projecting coast. When such a fall takes place, the most splendid and rarest fossils are brought to light, and made accessible almost without labour on their part. In the course of the past winter there had been no very favourable slips; the stock of fossils on hand was therefore smaller than usual.

Carus searched around until he found something that he was unable to resist purchasing:

still I found in the shop a large slab of blackish clay, in which a perfect Ichthyosaurus of at least six feet was embedded. The specimen would have been a great acquisition for many of the cabinets of Natural History on the Continent, and I consider the price demanded – £15 sterling [about £600 today] – as very moderate.

Either the doctor or the King bought the ichthyosaur at this very low price, and Carus added:

I was anxious at all events to write down the address, and the woman who kept the shop, for it was a woman who had devoted herself to this scientific pursuit, with a firm hand wrote her name 'Mary Annins' [sic] in my pocket-book, and added, as she returned the book into my hands, 'I am well known throughout the whole of Europe.'[51]

One must have, at the very least, a grudging admiration for this unusual and independent woman born into a strictly stratified society into which she did not fit. She seems to have spent the whole of her life in Lyme Regis (with the exception of one brief visit to London), dedicating herself to collecting fossils, analysing them, and acquiring a vast knowledge on the subject plus, of course, continuing to sell them in her shop.

She sold her last recorded but unspecified fossils to Sedgwick in October 1836. Her final sale to the British Museum (Natural History) was in 1840; it was of little-known ophiuroids (starfish), price paid £2 (today about £61), but at least this time her name was recorded, as the museum had had a change of policy in 1837.[52]

*   *   *

Latin is sometimes seen as a dead language for dead creatures – but it is for living ones too. Its inflexibility makes it ideal for universal scientific terminology, in order to identify consistently every living thing that is known. All the men who touched Mary's circle were honoured by having their name become part of the official Latin composite name of one or more living creatures. The popularity of geology meant that the public became much more aware of this use of Latin terminology. In a press report on a geological meeting in June 1855, which was attended by Murchison, the reporting journalist complained at length about the use of Latin, 'this very crabbed nomenclature which we think unnecessarily strews the road of such sciences with stumbling blocks', adding that the 'unintelligible gibberish [is] a grave difficulty to be surmounted'. He then urged that someone find a way 'more intelligibly descriptive than the fine Greek and Latin terms which are so constantly scraped from the heap of those dead languages'. This plea fell on deaf ears, and Latin's very inflexibility has ensured that it remains the universal scientific language.

All the puff and praise were for the gentlemen collectors and 'their' finds. But behind very many of the discoveries was a hard-working, largely unrecognised woman. To them, she was simply the commercial (therefore unworthy) source of specimens of vertebrate fossils for palaeontologists.

Genera or species were named after Murchison, including *Didymograptus murchisoni, Murchisonia bilineata* and *Megalaspidella murchisonae*, as were mountains, towns and rivers all over the world.

The fossil crab *Coleia antiqua* was named by naturalist and founder of the Zoological Society of London, William John Broderip, in acknowledgement of the collector William Willoughby Cole, Lord Cole (Earl of Enniskillen), and a characteristic local ammonite was named after Colonel Birch.[53] Even the Lord of the Manor at Lyme, Henry Hoste Henley, to whom Mary's first major find – the ichthyosaur – had been sold, was so honoured. Illustrator James Sowerby, when compiling his catalogue of fossil shells, named one species of ammonite after Gideon Mantell as a thank you for the perfect specimens Mantell had sent to him: *Ammonites mantelli*. And so it went on.

Females were not totally excluded. In fact, fossils were named in honour of two women in Mary's circle: for Miss

Elizabeth Philpot, it was a species of fossil fish, *Eugnathus philpotae*, so named in 1844 by the Swiss Louis Agassiz. He added, 'In dedicating this species to Miss Philpot, I have pleasure in publicly recognizing the services she has made to palaeontology and notably to fossil ichthyology, by the care she has taken in collecting the fossil remains of the Lias at Lyme Regis. The species we have just described and which is in her collection can be regarded as one of the finest fishes of this formation.'[54]

The second woman to be honoured in this way was Charlotte, Lady Murchison. Her own fine collection of specimens was used both by William Buckland and by James Sowerby, who described many of them in his book *Mineral Conchology of Great Britain* (1817). Sowerby named the cephalopod ammonite 'as a just tribute for the ardour with which she [Charlotte Murchison] has pursued the study of fossil conchology, the pleasing effects of which those who are so happy as to be acquainted with her know how to appreciate'.[55]

But such an honour was not for Mary. Mary's own finds were always named after someone else. William Buckland had many species named after him, and understood very well the importance of the official name; at one point he urged Conybeare to buy a certain specimen and then donate it to a museum to ensure that his name became part of its history forever. But even Buckland, Mary's great friend – whom she allowed to announce all her great finds – named the important *Plesiosaurus* she had found *P. macrocephalus*. And, for example, the starfish *Ophioderma egertoni*, which she had unearthed, was named by Professor Sedgwick after collector Sir Philip Egerton.

Unlike Elizabeth Philpot and Charlotte Murchison, Mary Anning was associated with the words 'woman', 'poor', 'lower class', 'in trade' and 'uneducated' (in the conventional sense). More tellingly, when so much, too much, is owed to one person, it is easier to ignore the point than begin to reveal the whole story. Those who had benefited so often may have been compelled to admit the long-running dishonour, the unmerited credit so often taken for her discoveries and her research. Yet these were exceptional men, in many ways honourable, who accomplished much, but lacked some essential decency by taking advantage of someone in no position to fight back. They were also reflecting the attitudes of their times.

It was the Swiss Louis Agassiz who finally established some justice for Mary, but not until 1841, when he named a fish, the species *Acrodus anningiae*, after her. He named another species of fish *Belenostomus [Belonorhychus] anningiae* in 1844. However, there were no such acknowledgements by British colleagues in her lifetime. According to Hugh S. Torrens, in 1878 R.F. Tomes named a new liassic coral genus and species *Tricycloseris anningi* after her, and in 1936 L.R. Cox named the bivalve genus *Anningia*, although this had to be changed in 1958 – because of a conflict in naming – to *Anningella*. Another tribute, less relevant to what she had accomplished, was the naming of the Karroo reptile genus *Anningia* in 1927, by the South African Robert Broom (1866–1951); in 1932 this became the core for a new order, the Anningiamorpa. Much more recently, in 1969, Alan Lord named a new ostracod species *Cytherelloidea anningi*.[56] Proper acknowledgement came slowly to the carpenter's daughter, but too late.

Mary continued to walk the cliffs and foreshore, as she had always done, continued her search for another great find, and as the years advanced her health began to fail.

6

The End and the Beginning of a Legend

Mary Anning [is] probably the most important unsung (or inadequately sung) collecting force in the history of palaeontology . . . She directly found, or pointed the way to, nearly every specimen of importance.

(Stephen Jay Gould[1])

Visitors to Lyme had increased over the years as Mary Anning won the respect of contemporary scientists. But the last decade of her life was a period of decline. It was all becoming more difficult. Fossils were no longer so rare or so novel, and prices fell dramatically, even for large specimens. There was less interest in fossils, and many more fossils were available. Important discoveries had evaded Mary for some time. She was forced to accept the low prices because of her poverty, and, unsurprisingly, she became disheartened. However, some of her colleagues remembered with kindness her accomplishments and their own debts to her. She received an income, and lived reasonably comfortably, through a combination of efforts by others on her behalf.

It began in May 1830 with the appearance of Henry De La Beche's illustration *'Duria antiquior'* or 'Ancient Dorset', almost bursting with reptilian life; this may have been inspired by Mary, even drawn for her benefit and converted into a saleable lithograph by George Scharf (1788–1860) to contribute to her financial support.[2] Those who saw it were thrilled to have sight of an image of the violent Jurassic sea in which one strange monster fought and ate another.

Then the British Association for the Advancement of Science, while meeting in Dublin in 1835, arranged for her to receive an annuity. It raised £200 for her through private subscription. The Government then boosted this fund with a donation of £300. Either William Buckland or Richard Owen had been successful in persuading Prime Minister William Lamb, Lord Melbourne (1779–1848), to approve it. This annuity began in 1838 at £25 a year (today about £815).[3] So in her later years she had enough to live on in her modest way.

To John Murray it was far from being enough. He later wrote, 'The pitiable pension doled out by the niggard hand of a Government . . . confines its honours and rewards to military prowess or naval heroism, and contrives to forget the imperishable triumphs of mind in science, literature and the arts . . . [This] reminds me of a similar trifle bestowed on the late Mary Anning, to whose contributions the science of geology is so largely indebted.'[4] In July 1846, the last year of her life, she became the first Honorary Member of the new Dorset County Museum in Dorchester, established only that year.[5] Financial support for Mary came from other sources as well. In 1846, after she had been diagnosed with cancer, the members at the Geological Society of London collected a stipend for her.[6]

At this late point in her life Mary had something in common with society women: the use of opium. It was a time when some ladies took a 'morning pip', which might be opium, to relieve the self-inflicted pain of unyielding corsets and the tight-lacing needed to achieve a wasp-like waist. It seems for Mary there was perhaps only laudanum, an opiate derivative, to help deaden her pain. The side effects can include sleepiness and/or euphoria, and with presumably increasing doses, some people thought she had taken to drink. Stoically, this remarkable woman carried on working as best she could to the end.

George Roberts, forever promoting Lyme Regis, claimed that 'persons in Lyme live to be a great age'. In 1833, when the population was around 3,400,[7] seventeen people were nearly 80, twenty-nine nearly 90 and one over 90. It was not to be for Mary.

Mary Anning had been born a Georgian, on 21 May 1799, during the reign of George III, and died a Victorian, when the still young Queen had been on the throne for the first ten of her sixty-four-year reign. Only three years after Mary Anning had been visited by the King of Saxony she departed this life. Her mother Molly had lived to be 78, but Mary had not reached 50 when she died of breast cancer, on Tuesday 9 March 1847. This disease is often thought to be a modern one but has been known since the time of the ancient Greeks. She had endured the pain of crab-like carcinoma of the breast for at least two years, at a time when there was no treatment or cure.

There had been at least one notable exception to this outcome: the famous diarist Fanny Burney, who had visited Lyme in 1791 on a health-seeking tour and 'rejoiced' when she left. In 1811, in France, she had a cancerous breast removed by Napoleon's Surgeon General. She survived to live for a further twenty-nine years, but her description of the operation on the kitchen table without anaesthetic remains harrowing even today.[8]

It had not been a long enough life for a strong, active woman who breathed fresh sea air every day of her life. Who can say why one person survives and another succumbs? The deprivations of poverty may have weakened her, as also the daily, anxious search for fossils to provide an income, and the ceaseless struggle against the elements. Then there was probably a lifetime of deep-seated and festering agitation and bitterness at not receiving true recognition for her work. All these must have taken their toll on her physical well-being.

In 1846 she had received a unique tribute, but, once again, it came too late. In the Minute Books of the Geological Society, the entry for 2 July reads: 'Resolved that Miss Mary Anning and [another] be requested to become Honorary Members of the Institution.' Official acceptance by the men with whom, in reality, she was at least an equal had come agonisingly slowly. Her name was printed in the list for that year.

Another unparalleled tribute came in the year after her death. Henry Thomas De La Beche had known her since boyhood. In his presidential address to the Geological Society, he celebrated the lives of those Fellows who had died in the past year, as was the custom. His old friend was included, even though she was not a Fellow. His words were as follows:

I cannot close this notice of our losses by death without adverting to that of one, who though not placed among even the easier classes of society, but one who had to earn her daily bread by her labour, yet contributed by her talents and untiring researches in no small degree to our knowledge of the great Enalio-Saurians, and other forms of organic life entombed in the vicinity of Lyme Regis . . . there are those among us in this room who know well how to appreciate the skill she employed (from her knowledge of the various works as they appeared on the subject) in developing the remains of the many fine skeletons of Ichthyosauri and Plesiosauri, which without her care would never have been presented to the comparative anatomists in the uninjured form so desirable for their examination. The talents and good conduct of Mary Anning made her many friends; she received a small sum of money for her services, at the intercession of a member of this Society with Lord Melbourne, when that nobleman was premier. This, with some additional aid, was expended upon an annuity, and with it, the kind assistance of friends at Lyme Regis, and some little aid derived from the sale of fossils, when her health permitted, she bore with fortitude the progress of a cancer on her breast, until she finally sunk beneath its ravages on the 9th of March 1847.[9]

This was the first time a tribute had been paid to a woman, and perhaps to a non-member; women were not accepted as Fellows of the Geological Society until 1904, almost sixty years after Mary Anning's death. In 1869 Girton College, Cambridge, accepted the first women at the university, and in 1874 came the first medical college for women in Britain. Did this speech prick any guilty consciences among those who had so freely used her work to make their own reputations and give themselves a generous income, without fully acknowledging their source?

Mary Anning's funeral was conducted by Revd Fred Parry-Hodges, whose sister had married into the Fanes, the family who held the patronage, that is, the control of the appointment, so their influence on Lyme continued to a degree. She was buried in the earth of the churchyard of St Michael the Archangel on Church Street over-looking Lyme. It was a place she knew well, and within sight of the bone-beds of the primeval creatures of her world. Her obituary was published in the *Quarterly Journal of the Geological Society*.

Mary 'freed' the 'dragons' but, equally, the 'dragons' 'freed' Mary. For her, it could easily have been a life as a mill worker or in service. There *was* a woman from a similar background whose life we can compare with Mary's. Ann Bennett was an infant girl born next door in Cockmoile Square in the same year as Mary. What kind of life did *she* have? Ann grew up to run a millinery and dressmaking business in Lyme's Broad Street with another woman, who married one of Ann's brothers. Ann herself at the late age of 33 married a widower, Richard Cox, who had a young son, and she moved to Bridport only 5 miles away. They produced three children, and Ann was to die of old age in the 1880s.[10]

In Lyme itself, Cockmoile Square was demolished in 1900, along with the small house the Annings had first lived in, in order to widen the still very narrow road. Among the shops swept away was a fossil shop, also named the Fossil Depot – a later enterprise than Mary

Anning's fossil shop – the shop of a fishmonger, Sidney Curtis, handily located between the bottom of Broad Street and the sea. Inside his shop, he sold freshly caught aquatic vertebrates, and – startlingly – examples of their fossilised ancestors from the depths of time. In his own way he was carrying on something of Mary Anning's tradition.[11]

The nearby Assembly Rooms building was sold and began a new life as a teashop and cinema before the First World War. It was in an exposed position, very close to the sea and on an unstable coast; unsurprisingly, the sea eventually won. Seriously damaged by a storm in 1927, it was demolished.

Most fittingly, a museum appeared on the very spot where Mary Anning's cottage and her Fossil Depot had stood along with two or three other cottages, although this was not realised until later. The Victorian Lyme Regis Philpot Museum was built by a great-nephew of the Philpot sisters. Thomas E.O. Philpot lived locally, was rich, and gambled. He was disliked, and may have built the museum to make himself more popular locally; however, when it was completed in 1901, there was little to put in the intended museum and art gallery. In 1921, one of Thomas's nieces, Miss Caroline Philpot, donated the building to Lyme. Finally set up as a trust in 1927, after a chequered history, the museum was re-established in 1960. Ever since, it has gone from strength to strength to become the perfectly located fine small museum that it is today.

Mary's second home in Broad Street became Beer and Son Hairdresser's and Tobacconists. The little thatched building is no longer there, but a plaque marks the spot.

Today, Lyme's fossil workshops give an idea of some of the work Mary did in preserving and preparing the sometimes very large fossils. Methods are not so very different today, although there are labour-saving electric tools, while Mary no doubt began with her father's carpenter's tools. But the prehistoric grey dust still clings to everything in the workshops of Lyme.

What happened to the Mary Anning collection and archive? No concentrated long-term effort was made to retain it as a whole. Lord Enniskillen acquired most of it in 1885 and, the year before he died, passed it on to Sir Richard Owen. Owen's archive was partly broken up – C.D. Sherborn selected the items he felt were important, and other artefacts were dispatched all over the world. Some items carry traces of their travels being stamped 'Coll. Sherborn. Ex Litt. Ricordi Owen. Don R.S. Owen.' No records were kept, so it is possible that letters and documents may still turn up.[12]

There were those among her colleagues who had always been quick to give her credit. Colonel Birch once wrote to Mantell that the Annings 'have in truth found almost all the fine things, which have been submitted to scientific investigation'. Mantell had said of William Bullock's Museum of Natural Curiosities: 'all the most valuable fossils had been obtained by the indefatigable labours of Miss Mary Anning.'[13]

In 1845 William Buckland became Dean of Westminster, having had his fill of the academic obstructionism to progress in science. The ever-present menagerie moved with the Bucklands; still part of the family was Tiglath, the bear – by now stuffed, as was Billy, the hyena. Buckland gradually gave up his trips to Oxford, where for so many years he had so amusingly lectured on the New Geology.[14] He focused his mind and energy wholly on fresh territory: Westminster Abbey and

Westminster School, with a view to reform and improve-
ment. This included emptying cesspits that had not been
cleaned for centuries, installing desperately needed new
sewers, and, in the face of the threat of cholera, constantly
promoting the theme 'Wash and be clean' to the
annoyance of some. Sadly, from at least 1850 Buckland's
mind became deranged, possibly from a fall, and he died
in 1856 in a mental asylum at Clapham. It was only three
years before Darwin's great opus was published. Mary
Buckland died the year after her husband.

Richard Owen's outstanding career had yet to peak in
1847. In 1836 he had been appointed Hunterian professor
of comparative anatomy and physiology at the Royal
College of Surgeons. Famously, he was the first to identify
the moa (after seeing only one bone), a giant, flightless bird
from New Zealand. In a report of 1842 on the British fossil
Reptilia, he gave the name *Dinosauria* to a group of large
terrestrial reptiles, for which he will always be remembered. Much more was to come, including the founding of
London's Natural History Museum. There, his statue
halfway up the great marble staircase gazes out on the
cathedral-like structure around him and down on the giant
figure of *Diplodocus* in the entrance hall. This dinosaur,
the longest land animal ever found, is itself always
surrounded by wide-eyed children and adults looking up.

Roderick Impey Murchison, the most decorated and
honoured of all the pioneer geologists, was knighted in
1846. He never fully recovered from the death of his wife,
Charlotte, in 1869. William Conybeare became totally
immersed in theological matters in the 1840s, and
deserted geology completely. At the time of Mary's death
he was still vicar at Axminster, but soon to leave; he
outlived her by only ten years. Henry De La Beche

received many honours in recognition of his accomplishments and administrative ability. In 1847, just before Mary's death, he was elected President of the Geological Society, and was also Director of the Geological Survey. Thomas Hawkins passed on in 1889.

In 1846, Louis Agassiz – his obsession with fossil fish having destroyed his marriage – visited the United States, where he remained to become one of the founding fathers of American scientific tradition.

Anna Maria Pinney, whose journal revealed some hidden facets of Mary Anning's personality, never married and departed this life when only young. Mary's friend Elizabeth Philpot, who was perhaps twenty years older than her, outlived her by ten years; her obituary read: 'she went upon the lias shore with Mary Anning almost daily.'[15]

A nostalgic working image of Mary and some of her distinguished colleagues remains. The Earl of Enniskillen (formerly Lord Cole), for whom Mary had supplied many specimens, wrote to Sir Richard Owen in 1885 – the last letter Owen received: 'neither of us, my dear friend, are as young as we were, or near as active as when we used to clamber over the cliffs with Mary Anning.'[16]

From the time she miraculously survived a lightning strike as an infant, events in Mary Anning's life would prove to be myth-making. Other elements in her story would add to the legend: her extreme youth when she made her first find, and the relatively little known about her. She wrote no scholarly articles, nor would she have been considered qualified to do so. Mary Anning's childhood discovery of the *Ichthyosaurus* probably outshone almost everything else she accomplished, although there was much more to come of geological and

palaeontological value. When there are gaps in knowledge, the vacuum can be filled with myth, especially in
reference to a woman, and an unusual woman at that.

To some the gentle maiden was – irresistibly – a slayer
of fearsome dragons. With the find of the winged *Pterodactylus macronyx* (renamed *Dimorphodon macronyx*),
which fascinated, thrilled and terrified the public, she
became 'St Georgina of Lyme Regis'.[17] From time
immemorial dragons have been part of universal human
mythology. It is not surprising that prehistory and myth
sometimes merged; especially with the discovery of
inexplicably enormous and strange fossils. In the early
days the giant marine reptiles, the ichthyosaurs and
plesiosaurs, were called 'sea dragons'. Dragons, feared in
the West, were, and are, revered in the East. For centuries
in China, prehistoric teeth and bones of unfathomable
size have been ground up to create potions to cure ailments, and it was Chinese apothecaries who led the first
fossil hunters to the bone-beds.

Mary Anning's uniqueness and her success in fossiling
drew effusive appellations. Thomas Hawkins and George
Roberts conjured up Greek mythology when speaking
about her. To historian Roberts, she was 'a Helen to the
geologists' – Helen being the most famous and most
beautiful woman of Greek mythology – 'the progressive
discovery of the structure of the Ichthyosaurus taking
about the same number of years as the siege of Troy'.[18]
It had been ten years from the first *Ichthyosaurus* find by
Mary to the establishment and adequate description of
the genus; over that long period of time, it was Mary who
worked hard in continuing to supply the learned
gentlemen with the fragments necessary to complete the
definitive specimen.

To Hawkins she was a Pythoness (prophetess). A German visitor called her the 'Princess of Palaeontology'. Even to critical Gideon Mantell she was 'the geological Lioness'. To the Victorians her story was an appealing folk tale at a time when it was essential that a young heroine be pretty, with long, loose hair, beribboned, wearing a sweet frock – a mere child making great discoveries, as Mary was depicted in Arthur Mee's encyclopedia.

She was famous enough, however, to be commemorated in a comical rhyme. It was a kind of fame, although it trivialises her achievements, and it is unlikely this would have been written about a man.

> Miss Anning, as a child, ne'er passed
> A pin upon the ground;
> But picked it up, and so at last
> An Ichthyosaurus found.[19]

Mary was neither rich, nor well travelled, nor from the monied middle and upper classes. She did not publish any learned papers. After all the great discoveries she had made, it fell to the Swiss Louis Agassiz, to name two British species after her. Verses were written celebrating her, but although sincere they were sadly lacking in distinction. Two years after her death Mary was joined in St Michael's churchyard by her older brother, Joseph, who had served as church warden from 1844 to 1846. Three of his infant children share the same grave. The following words appear on their gravestone:

Sacred to the memory of Joseph Anning who died July 5th, 1849, aged 54 years. Also of three children who died in their infancy. Also of Mary Anning sister of the above, who died March 9th 1847, aged 47 years.

The Revd Fred Parry-Hodges, an Anglican with an evangelical approach and a fossil collector himself, and friends arranged for a stained-glass window by William Wailes (1808–188?) to be placed in St Michael's in Mary Anning's memory. Wailes's glass was noted for its rich colours and medieval style. A.W.N. Pugin, who designed the Parliament buildings, used Wailes's glass exclusively from 1842, and the firm would later exhibit at the Great Exhibition of 1851. The window was in place in February 1850. The stained-glass pictures depict 'the six corporeal acts of mercy': visiting the sick, feeding the hungry, giving drink to the thirsty, giving clothing for the naked, sheltering the homeless and visiting orphans. This theme was particularly apt for Mary Anning. It was also probably much less expensive than a bespoke window designed with her image on it. The legend below the window reads:

> This window is sacred to Mary Anning, of this parish, who died A.D. 1847, and is erected by the Vicar of Lyme and some of the members of the Geological Society of London, in commemoration of her usefulness in furthering the science of geology, as also of her benevolence of heart and integrity of life.

As interest waned, some continued to keep watch over her memory. In the early twentieth century Mr W.J. Harding suggested that lettering on her gravestone should be renewed. The money to do this was quickly raised by geologists and others. In 1929 naturalist Gulielma Lister of High Cliffe, and her sister, asked the vicar for permission to renew the legend on the stained-glass window, which they did with oil paint: 'We felt it a privilege to be allowed to do this.'[20]

To mark the bicentenary of Mary's birth an international symposium was held at the Lyme Regis Philpot Museum from 2 to 4 June 1999, entitled 'Mary Anning and Her Times: The Discovery of British Palaeontology, 1820–1850'. The symposium was convened by Mary Anning's great-great-great-nephew, Sir Crispin Tickell, a distinguished diplomat, scholar and environmentalist. Through him we know that the Anning line of strong women continued after Mary; one of his aunts was secretary to Sir Winston Churchill, while another aunt became secretary to the campaigning journalist, W.T. Stead. At the gathering to honour Mary Anning's life, there were keynote lectures by Hugh S. Torrens, Stephen Jay Gould, Sir Crispin Tickell and John Fowles.

In 2001 around 95 miles of the coastline, running from Exmouth in East Devon to Old Harry Rocks near Swanage in West Dorset (with Lyme Regis right in the middle), were declared England's first natural World Heritage Site, giving it equal stature to the Grand Canyon and the Great Barrier Reef. It is 'An outstanding example representing major stages of the Earth's history, including the record of life, significant ongoing geological processes in the development of landforms, and significant geomorphic or physiographic features' (UNESCO declaration of 13 December 2001).

In Lyme Regis, there is an Anning Road and, on the site of her first home, the Lyme Regis Philpot Museum features displays on her and her work, and an annual 'Mary Anning Day' celebrates her life and achievements. The true tribute comes from the thousands of visitors who seek the fossils from another world – the world of 'the greatest fossilist the world ever knew'.[21] Mary Anning was not a village blue-stocking, not a dainty romantic, not a feminist, not an adventuress. She was herself – tough, strong, humble, plain, practical, honest, curious, intelligent and with intellectual mastery of her subject. She was a woman today's woman can appreciate. Of all the tributes to her character, poet John Kenyon sums her up best: 'those who knew her personally, will be no less eager to bear testimony to her kindly temper, her straight-forward character, and her fresh and vigorous intellect.'[22]

In 1799, when Mary Anning was humbly born to a poor carpenter, geology was not a science, nor was palaeontology. No one knew what the strange objects found in the cliffs were, or how they had got there. The story of Creation was interpreted literally. There was no geological timescale and there were no geological Periods. Dinosaurs had not yet been discovered, described and named. By the time she died in 1847, there were Professors of Geology at Oxford and Cambridge. Most geologists were at least willing to consider that Creation – with man as the exception – meant that the whole of the natural universe owed its structure to an evolutionary plan, not to dramatic volcanic eruptions or to the Flood.

Credit for this was due to a number of exceptional, mainly privileged, people, most within Mary Anning's circle of acquaintance. But Mary's largely unheralded contribution was as great as anyone's: she provided much of the raw material. Bearing in mind everything against her – sex, poverty, class, religion, lack of education, unmarried status, location in rural Dorset – whose contribution was the greater? In the end, does it matter who led the way, who discovered what and when? The answer is yes it does. The afterglow of Mary Anning's life can

still be felt. In Lyme Regis, during her lifetime she was a
boon to the tourist industry. She still is. Visitors connect
with the still evocative atmosphere and surroundings in
which she made her great discoveries, and search for
fossil treasures on the foreshore themselves in this
lovely setting.

In keeping with Mary Anning's private, independent
spirit, she retains her mystery to this day – all part of the
story of a remarkable woman whose flame burns ever
brighter in Lyme Regis and in the annals of the sciences
of geology and palaeontology.

Appendix
Mary Anning's Fossils

HER FIVE GREAT DISCOVERIES

1. 1812: The first complete skeleton of an *Ichthyosaurus* ('fish lizard') (renamed *Temnodontosaurus platyodon*). Mary's brother Joseph found the head in 1810/11 and Mary unearthed the body one year later. Sold to Henry Hoste Henley, who sold it on to William Bullock's Museum of Natural Curiosities in London; in 1820 it went to the Hawkins Collection, then the skull was passed to the Natural History Museum. Today: only the skull is in the museum (Ref. No. 1158).

2. December 1823: *Plesiosaurus giganteus* became the type fossil. Found under Black Ven cliff. Announced in January 1824 by Mary Anning in letter to Buckland, and she included a 'rude sketch' of it in ink. Conybeare and De La Beche had already done much work on this creature before a complete specimen was found; Conybeare took the credit. Purchased by Duke of Buckingham. Today: Natural History Museum (Ref. No. 22656).

3. December 1828: *Pterodactylus macronyx* (now *Dimorphodon macronyx*): the first British discovery of this flying dragon. Initially Buckland purchased it to save it. Today: Natural History Museum has three bones (Reg. No. R 1034); not on display. Teeth (Philpot collection) Oxford University Museum (Reg. No. J.28251).

4. December 1829: *Squaloraja polyspondyla* Agassiz, the type specimen. New fossil fish, a chimaera at the time incorrectly believed to be a transitional link between sharks and rays. Went to the Bristol Institution for the Advancement of Science, Literature and the Arts, and destroyed in Second World War. Today: tail (Philpot collection) Oxford University Museum (Reg. No. J.3097).

5. 21 December 1830: *Plesiosaurus macrocephalus*. Fossilised in a curled-up position. Announced in letter from Mary Anning to Buckland, 21 December, describing it as 'Most beautiful'. Named by Buckland 1836; described by Owen in 1840. Bought by Lord Cole (Enniskillen) for 200 guineas. In 1883 the Enniskillen collection went to Natural History Museum. Today: Natural History Museum (Enniskillen 1883, Reg. No. R 1336).

Additionally, she identified coprolites in 1824 and *Belemnosepia* in 1826.

In *Roberts's History of Lyme Regis and Charmouth* (1834), George Roberts listed his version of the organic remains found by Mary Anning. Some spellings and names have changed since Roberts's day. Modern names or further information are in square brackets:

Organic Remains of the Lias

REPTILLA
1. *Pterodactylus macronyx* [renamed *Dimorphodon macronyx*] Buckland, Crocodilus?
1. *Plesiosaurus dolichodeirus*, Conybeare.
2. —— *macrocephalus*, Conybeare.

1. *Ichthyosaurus communis*, De La Beche.
2. —— *platyodon*, De La Beche [Conybeare].
3. —— *tenuirostris*, De La Beche [Conybeare].
4. —— *intermedius*, Conybeare.

PISCES
1. *Dapedium politum*, De La Beche.
Several undescribed species of fish.
Ichthyodorulites [fin spines of *Hybodus*, a primitive shark], Buckland and De La Beche, several kinds.
Fish, teeth, palates, &c.

CRUSTACEA
Several species not yet described.

MOLLUSCA
Rhyncolites, or Sepia Beaks.
1. *Ammonites heterophyllus*, Sowerby.
2. —— *Henleii*, Sowerby.
3. —— *communis*, Sowerby.
4. —— *fimbriatus*, Sowerby. &c.[1]

Mary Anning was alive when Roberts wrote this book; indeed she was a subscriber to it. She must have provided the list or at least seen it, yet it is *incomplete*.

Oddly, the fossil fish *Squaloraja*, her fifth great find, is not mentioned, nor is brittle star, *Ophioderma egertoni*, or the ink sacs of belemnites or her work on coprolites.

She found several specimens of the large reptiles, including *Ichthyosaurus platyodon*, renamed *Temno-dontosaurus platyodon* (the largest ever found), *Ichthyo-saurus vulgaris*, *Ichthyosaurus intermedius*, *Ichthyosaurus communis* and *Plesiosaurus dolichodeirus*. She did much work on coprolites with Buckland, and on *Belemnosepia*: excellent invertebrate cephalopod specimen with its fossil ink in ink sacs; and *Apylsia punctata* (sea hare). Her correspondence with Professor Sedgwick reveals that he acquired at least seven specimens from her, but only one can be positively confirmed: the *Ichthyosaurus* (SM J.352189), which was sketched by her in her letter to him of 4 May 1843.

It seems that most of the fossils Mary Anning found were not credited to her. At that time the commercial supplier was not acknowledged. Her specimens were often, of financial necessity, sold to institutions and private collectors, whose records noted only the people who had donated the fossils to an institution, usually those from the moneyed and upper classes. There are, for example, several 'probables' at the Sedgwick Museum. This has made it difficult for researchers to trace the numerous fossils that Mary Anning unquestionably unearthed.

Some natural history museums are re-examining their documents and collections in an attempt to confirm which specimens were found by Mary. From this distance in time, although more research needs to be done, it may never be possible to be certain.

WHERE TO SEE EXHIBITIONS ON MARY ANNING AND HER FOSSILS

Bath Royal Literary and Scientific Institution

Stone lily, *Pentacrinus briareus*
Ichthyosaurus communis

Dinosaur Museum, Dorchester

'Mary Anning Room' on her life, with a realistic statue

Dorset County Museum, Dorchester

Exhibition on Mary Anning

Geological Society, London

Portrait in pastels of Mary Anning by B.J.M. Donne

Lyme Regis Philpot Museum

Mary Anning, 'our townswoman', features in many displays. To be seen are Mary's fossil extractor, a polished stone table top, comprising slices of coprolites, made for Buckland, and one of the Philpots' 'museum' cabinets. In 1999 the museum was named South West Museum of the year, when David Attenborough described it as 'a delightful and remarkable museum, a jewel'.

Muséum National d'Histoire Naturelle, Paris

Specimens purchased for Cuvier at Bullock's sale (1820)
Plesiosaurus dolichodeirus, purchased from Mary in
  1824

Natural History Museum

Ichthyosaurus platyodon, unnamed
Skull of *Temnodontosaurus platyodon* (found by Joseph
   Anning, 1810/11) (Reg. No. R 1158)
Plesiosaur (first complete, found 1823) (Reg. No. 22656)
Flying pterosaur, *Dimorphodon macronyx*, fragments (not
   on display) (found Dec. 1828) (Reg. No. 1034)
Ancient crab, *Coleia antiqua*
Plesiosaurus dolichodeirus (second complete, found
   1829) (Reg. No. 1313)
Ophiuroids, three (purchased from Mary, 1840) (Reg. Nos
   14398–400)
Coprolites, two
Plesiosaurus macrocephalus (second complete, found
   Dec. 1830) (Reg. No. 1336)
Ichthyosaur, *Leptopterygius platyodon*, incomplete
   (found 1832)
Probably many other specimens from the Willoughby/
   Cole/Enniskillen collection, much of which originated
   with Mary; also Hawkins's collections (purchased
   1834, 1840), including the massive *Temnodontosaurus
   platyodon*
Portrait in oils of Mary Anning, by William Gray

Oxford University Museum of Natural History

Philpot Collection: *Plesiosaurus dolichodeirus*, among
   many other unidentified Anning specimens;
Ichthyosaurus communis with stomach contents (found
   1835), donated by Lord Cole (Reg. No. J.13587);
Dimorphodon macronym (teeth) (Reg. No. J.28291)
Squaloraja (tail) (Reg. No. J.3097)

Belemnite: anterior sheath and ink sac (found 1828) (Reg. No. J.03572)

Belemnite: phragmacone (internal cone), purchased from Mary (Reg. No. J.03578)

Coprolite, 4cm high, presented by Mary (Reg. No. 23781)

Sedgwick Museum of Earth Sciences, Cambridge

Exhibition: 'Life in the Jurassic Sea'

Ichthyosaurus (SM J.35190) (definite)

Ichthyosaurus head, with six curved vertebrae attached (SM J.59645) (probable)

Ichthyosaurus post-cranial skeleton (SM J.59642) (probable)

Ichthyosaurus communis, with 5-foot-long skeleton (probable)

Ichthyosaurus (SM J.35187) (probable)

Ichthyosaurus platyodon (SM J.68446) (probable)

Ichthyosaurus (SM J.35189)

Stone lily, *Pentacrinus briareus*, purchased from Mary

Crinoid-bearing slab

Pentacrinite (purchased 1820)

Oil painting, *A Victorian View of Life in the Jurassic, c. 1850*, by Robert Farren (1832–1910), based on the famous sketch *Duria antiquior*, drawn by Henry De la Beche in 1831.

Somerset County Museum, Taunton

Ammonites oxynoticeras (found 1838)

Ulster Museum, Belfast

Ichthyosaur (Reg. No. BELUM K1296)[2]

Notes

Chapter One

1. G.A. Mantell, *Thoughts on a Pebble, or A First Lesson in Geology*, London, Reeve, Benham and Reeve, 1849.
2. The precise date is unclear. It was either 1810 or 1811, but probably 1811, because in November 1812 the *Western Flying Post* reported the extraction of the entire skeleton, which was one year after the skull was found.
3. *Western Flying Post*, November 1812.
4. Journals of Anna Maria Pinney 1831–3, quoted in W.D. Lang, 'Mary Anning and Anna Maria Pinney', *Proceedings of the Dorset Natural History and Archaeological Society*, 76 (1956), pp. 146–52.
5. George Roberts, *Roberts's History of Lyme Regis and Charmouth*, London, Samuel Bagster and William Pickering, 1834, p. 11.
6. Hugh Torrens, 'Presidential Address: Mary Anning (1799–1847) of Lyme; "The Greatest Fossilist the World Ever Knew"', *British Journal of the History of Science*, 28 (1995), p. 258.
7. Jill Warner and Pam Bennett Gupta, *The Bennetts of Lyme Regis 1762–1911*, Wimborne, Dovecote Press, p. 22.
8. The Marine Theatre is on the site today.
9. Roberts, *History of Lyme Regis*, pp. 286–7.
10. Hugh S. Torrens, 'Mary Anning's Life and Times: New Perspectives', Mary Anning Symposium, Lyme Regis Philpot Museum, June 1999.

11. G. Roberts, *A Social History of the People of the Southern Counties of England in Past Centuries*, London, Longman, Brown, Green, Longman & Roberts, 1856, p. 562; Roberts, *History of Lyme Regis*, p. 287.
12. Roberts, *History of Lyme Regis*, pp. 286–7.
13. Where the public toilets are today.
14. Now the Dinosaur Museum.
15. Cyril Wanklyn, *Lyme Regis: A Retrospect*, London, Arthur L. Humphreys, 1922, p. 70.
16. Roberts, *Social History*, p. 559.
17. The route is no longer there; the sea took it away just after Mary Anning died in 1847.
18. Jane Austen, *Persuasion*, London, Penguin, 1998, p. 89.
19. Hugh S. Torrens, Mary Anning Symposium, quoted in Thomas Goodhue, *Fossil Hunter: The Life and Times of Mary Anning, 1799–1847*, Bethesda, Md, Academica Press, 2004, p. 11.
20. Roberts, *Social History*, p. 557.
21. Now in the Lyme Regis Philpot Museum.
22. Cunnington MSS, Wiltshire Archaeological Society Library, Devizes, quoted in Torrens, 'Presidential Address', p. 259.
23. J.A. De Luc, *Geological Travels*, vol. II: *Travels in England*, translated from the French manuscript, 1811, quoted in W.D. Lang, 'Mary Anning and the Pioneer Geologists of Lyme', *Proceedings of the Dorset Natural History and Archaeological Society*, 60 (1939), pp. 145, 147.
24. Lang, 'Mary Anning and the Pioneer Geologists', p. 145.
25. De Luc, *Geological Travels*, quoted in Lang, 'Mary Anning and the Pioneer Geologists', pp. 145–6.
26. Roberts, *History of Lyme Regis*, p. 288.
27. Torrens, 'Presidential Address', p. 259.
28. Roland Brown, *The Beauties of Lyme Regis, Charmouth, the Landslip and their Vicinities*, London, Longman, Brown, Green & Longmans, 1857, p. 26.
29. Roberts, *History of Lyme Regis*, p. 288.

30. Lang, 'Mary Anning and Anna Maria Pinney', pp. 146–52.
31. Elizabeth Oke Buckland Gordon, *The Life and Correspondence of William Buckland, DD, FRS, Sometime Dean of Westminster, Twice President of the Geological Society, and First President of the British Association*, London, John Murray, 1894, p. 1.
32. *Ibid.*, pp. 2–4.
33. *Ibid.*, p. 3.
34. *Ibid.*, p. 114.
35. *Chambers's Journal* (1858), p. 314, quoted in W.D. Lang, 'More about Mary Anning, Including a Newly-Found Letter', *Proceedings of the Dorset Natural History and Archaeological Society*, 71 (1950), p. 187.
36. H. De La Beche and W.D. Conybeare, 'Notice of the Discovery of a New Fossil Animal', *Transactions of the Geological Society of London*, 1821, 5, pp. 559–94, plates 40–2.
37. Torrens, 'Mary Anning's Life and Times: New Perspectives', Mary Anning Symposium, June 1999.
38. Gideon Mantell MSS, Alexander Turnbull Library, Wellington, New Zealand, quoted in Torrens, 'Presidential Address', p. 261.
39. Roberts, *History of Lyme Regis*, p. 324.
40. *Ibid.*
41. Torrens, 'Presidential Address', p. 261.
42. British Library, Add. MS 36520, quoted in Torrens, 'Presidential Address', p. 261 (emphasis in original).
43. Torrens, 'Presidential Address', p. 262.
44. W.D.I. Rolffe, A.C. Milner and F.G. Hay, 'The Price of Fossils', *Special Papers in Palaeontology*, 40 (1988), p. 149, quoted in Torrens, 'Presidential Address', p. 262.
45. *Ibid.*
46. British Museum archives, letter bound in C 1467 (Torrens noted wrongly dated 1820); Torrens, 'Presidential Address', p. 262. This and other specimens in the Bristol Institution

that had been found by Mary Anning were destroyed by bombing on the night of 24/25 November 1940 in the Second World War. The tail section survives at Oxford University Museum.

47. *Bristol Mirror*, 11 January 1823, p. 4, quoted in Torrens, 'Presidential Address', p. 263.

48. Roberts, *History of Lyme Regis*, pp. 162, 163.

49. W.D. Lang, 'Three Letters of Mary Anning, "Fossilist" of Lyme', *Proceedings of the Dorset Natural History and Archaeological Society*, 66 (1945), p. 169.

50. Roberts, *Social History*, p. 553.

51. Roberts, *History of Lyme Regis*, p. 163.

52. Wanklyn, *Lyme Regis: A Retrospect*, p. 85.

53. *Ibid.*, p. 2.

54. Where the car park now stands.

55. Constance Hill, *Austen: Her Homes and her Friends*, London, Bodley Head, 1901, pp. 143–4.

56. John Fowles, *A Short History of Lyme Regis*, Wimborne, Dovecote Press, 1991, p. 33.

57. Jane Austen, *Persuasion*, p. 89.

58. The later supposition that the heroine fell on 'the steep flight of steps' known as 'Granny's teeth' was wrong since they date from later – 1825, when Mary Anning was a young woman.

59. Jane Austen, Letter 39 to Cassandra Austen, in *Jane Austen's Letters*, ed. Deidre Le Faye, Oxford, Oxford University Press, 1995.

60. C.G. Carus, *The King of Saxony's Journey through England and Scotland in the year 1844*, London, Chapman & Hall, 1846, p. 197.

61. Gordon, *Life and Correspondence of William Buckland*, p. 115.

62. Roberts, *Social History*, pp. 556–7.

Chapter Two

1. Hutton to George Clerk Maxwell, August 1774, National Archives Scotland, GD 18/5937/2.
2. J. Playfair, 'Biographical Account of the late James Hutton, FRS Edinburgh', *Transactions of the Royal Society of Edinburgh*, 5/3 (1805), p. 56.
3. James Hutton, *Theory of the Earth with Proofs and Illustrations*, Edinburgh, Messrs Cadwell, Junior, Davies and W. Creech, 1795, vol. 1, p. 304.
4. Gordon, *Life and Correspondence of William Buckland*, p. 19.
5. Playfair, 'Biographical Account', p. 73.
6. George Cumberland, Letter to Editor, *Monthly Magazine*, 1815, quoted in Deborah Cadbury, *The Dinosaur Hunters*, London, Fourth Estate, 2000, p. 21.
7. Roberts, *Social History*, p. 557.
8. Gideon Mantell, *The Journal of Gideon Mantell, Covering the Years 1818–1852*, ed. E.C. Curwen, Oxford, Oxford University Press, 1940, p. 108.
9. W.D. Lang, 'Mary Anning and a Very Small Boy', *Proceedings of the Dorset Natural History and Archaeological Society*, 84 (1963), p. 181.
10. After his death, these specimens and labels were given to the museum at Weston-super-Mare, and were lost.
11. Lang, 'Mary Anning and a Very Small Boy', p. 182.
12. W.D. Lang, 'Mary Anning and the fire at Lyme in 1844', *Proceedings of the Dorset Natural History and Archaeological Society*, 74 (1953), p. 175.
13. Richard Owen correspondence, Earth Sciences Library, Natural History Museum, vol. 1, pp. 149–50.
14. Gordon, *Life and Correspondence of William Buckland*, p. 194.
15. *Ibid.*, p. 27.
16. *Ibid.*, p. 36.
17. *Ibid.*, p. 24.

18. *Ibid.*, p. 10.
19. *Ibid.*
20. Cadbury, *The Dinosaur Hunters*, p. 61.
21. Gordon, *Life and Correspondence of William Buckland*, pp. 104–5.
22. Fowles, *A Short History of Lyme Regis*, p. 40.
23. Edwin Colbert, *Man and Dinosaurs*, New York, Penguin, 1968, quoted in Cadbury, *The Dinosaur Hunters*, pp. 59–61.
24. Philip Duncan, 'A Picture of the Comforts of a Professor's Rooms in CCC, Oxford', 1821, in C.G. Dauberg, *Fugitive Poems Connected with Natural History and Physical Science*, Oxford, Parker and Sons, 1861, quoted in Gordon, *Life and Correspondence of William Buckland*, p. 9.
25. Gordon, *Life and Correspondence of William Buckland*, p. 113.
26. J.M. Edmonds, 'The Fossil Collection of the Misses Philpot of Lyme Regis', *Proceedings of the Dorset Natural History and Archaeological Society*, 98 (1976), pp. 44–5.
27. Quoted in Cadbury, *The Dinosaur Hunters*, p. 256.
28. Mantell, *Journal*, p. 108.
29. Now the site of Lloyd's Bank.
30. Conybeare, letter to the President of the Geological Society, W.J. Hamilton, 18 April 1855.
31. Paul J. McCartney, *Henry De La Beche: A New Kind of Geologist*, Cardiff, Friends of the National Museum of Wales, 1977, p. 28.
32. *Ibid.*, p. 29.
33. *Ibid.*, p. 31.
34. Roberts, *History of Lyme Regis*, p. 221.
35. Thomas W. Goodhue, 'The Faith of a Fossilist: Mary Anning', *Anglican and Episcopal History*, 70/1 (2001), p. 92, quoting W. Densham and J. Ogle, *The Story of the Congregational Churches of Dorset*, Bournemouth, 1899, p. 152, and Revd S. Knell, 'Church Book', manuscript, Lyme Regis, Coombe Street Chapel, 1852–6.

36. J. Murray, 'The Late Miss Mary Anning', *Mining Journal*, 17 (25 December 1847).
37. Lang, 'Mary Anning and Anna Maria Pinney', pp. 146–52.
38. Edmonds, 'Fossil Collection of the Misses Philpot', p. 43.
39. One of two cottages that formed the later inn, now part of the Mariners Hotel. Beatrix Potter visited Lyme in 1904 and lodged in Silver Street. By her own account, in *Little Pig Robinson* (1930) the illustration of the coaching inn was loosely based on the Mariners Hotel (as it looked at that time). It was no accident that she was in Lyme – she too had been fascinated by fossils since childhood.
40. Thomas Allan, 'Travels in England 1813–1824', London, Palaeontological Library, Natural History Museum, quoted in Lang, 'Collector and Vendor of Fossils, 1799–1837', *Natural History Magazine*, 77.
41. Edmonds, 'Fossil Collection of the Misses Philpot', p. 43.
42. *Ibid.*, p. 47.
43. Now a hotel in the Sidmouth Road.
44. 'Lister's Thesaurus', p. 30B. MS compiled by Miss Gulielma Lister, helped by Miss Blanche Palmer. Cyril Wanklyn gave it the title 'Recollections of Lyme'. Miss Lister donated it to the Lyme Regis Philpot Museum.
45. Edmonds, 'Fossil Collection of the Misses Philpot', p. 47.
46. Lang, 'Three Letters', p. 169.
47. All Bell references, list and exchange of letters from Grant Johnson, *A Memoir of Miss Frances Augusta Bell*, London, Hatchard & Son, 1827.
48. Lang, 'Mary Anning and Anna Maria Pinney', p. 147.
49. Interleaved copy of Roberts, *History of Lyme Regis*, quoted in W.D. Lang, 'Portraits of Mary Anning and Other Items', *Proceedings of the Dorset Natural History and Archaeological Society*, 81 (1959), pp. 89–91; and Lang, 'More about Mary Anning', 71 (1949), p. 187.
50. Roderick Gordon, private collection.

51. *Chambers's Journal* (1857), p. 314, quoted in Lang, 'More about Mary Anning', p. 187.

Chapter Three
  1. Roberts, *History of Lyme Regis*, pp. 325–6.
  2. *Western Flying Post*, 15 December 1823, quoted in Torrens, 'Presidential Address', p. 263.
  3. George Roberts, *An Etymological and Explanatory Dictionary of the Terms and Language of Geology*, London, Longman, Orne, Brown, Green & Longmans, 1839, p. 136.
  4. William Buckland, 'On the Discovery of a New Species of Pterodactyle in the Lias at Lyme Regis', *Transactions of the Geological Society of London*, 2nd series, 3 (1829), pp. 217–22.
  5. Thomas Hawkins, *Memoirs of Ichthyosauri and Plesiosauri: Extinct Monsters of the Ancient Earth with Twenty-One Plates Copied from Specimens in the Author's Collection*, London, Relfe & Fletcher, 1834, p. 39.
  6. Lang, 'Mary Anning and the Pioneer Geologists', pp. 152–3.
  7. *Ibid.*, p. 151.
  8. *Ibid.*, p.153.
  9. Torrens, regarding unpublished letter from Warburton to Buckland, quoted in Cadbury, *The Dinosaur Hunters*, p. 110.
 10. Gordon, *Life and Correspondence of William Buckland*, pp. 28–30.
 11. *Ibid.*, pp. 21–2.
 12. *Ibid.*, p. 103.
 13. *Ibid.*, p. 13.
 14. Thomas Allan, 'Travels in England', unpublished journal, quoted in Lang, 'Mary Anning and the Pioneer Geologists', p. 154.
 15. All Bell references, list and exchange of letters from Johnson, *A Memoir of Miss Frances Augusta Bell*.
 16. Goodhue, 'Faith of a Fossilist', p. 96.

17. Michael A. Taylor, 'The Lyme Regis (Philpot) Museum, The History, Problems and Prospects of a Small Museum and its Geological Collection', *Geological Curator*, 4/6 (1986), p. 301.

18. Probably J.35187.

19. David Price, 'Mary Anning Specimens in the Sedgwick Museum, Cambridge', *Geological Curator*, 4/6 (1986), p. 321.

20. Roberts, *History of Lyme Regis*, p. 174.

21. Warner and Gupta, *Bennetts of Lyme Regis*, p. 41–2.

22. Gordon, *Life and Correspondence of William Buckland*, p. 113.

23. Lang, 'Mary Anning and the Pioneer Geologists', p. 55.

24. Maria Hack, *Geological Sketches and Glimpses of the Ancient Earth*, London, Harvey & Dutton, 1832, p. 302.

25. Lang, 'Mary Anning and Anna Maria Pinney', pp. 146–52.

26. E.B. and D.S. Berkeley, *George William Featherstonhaugh, the First US Government Geologist*, Tuscaloosa, Ala., University of Alabama, 1988, quoted in Torrens, 'Presidential Address', p. 265.

27. Hugh Torrens and M. Taylor, 'Saleswoman to a New Science: Mary Anning and the Fossil Fish Squaloraja from the Lias of Lyme Regis', *Proceedings of the Dorset Natural History and Archaeological Society*, 108 (1986), p. 135.

28. Nellie Waring (Sister Emma) (S.E.), *Peeps into an Old Playground: Memoirs of the Past*, Lyme Regis, Dunster, 1895, p. 4.

29. 'X', *Lyme 50 Years Ago*, Lyme Regis, 1892.

30. *The Journal of Gideon Mantell Surgeon and Geologist: Covering the Years 1818–1852*, ed. E. Cecil Curwen, Oxford, Oxford University Press, 1940, p. 108.

31. Gordon, *Life and Correspondence of William Buckland*, pp. 7, 8.

32. Edmonds, 'Fossil Collection of the Misses Philpot', p. 45.

33. Lang, 'Mary Anning and Anna Maria Pinney', p. 149.

34. Murray, 'Late Miss Mary Anning', p. 591.
35. William Buckland, 'Fossil Sepia', *London and Edinburgh Philosophical Magazine*, 5 (1829), p. 388.
36. Edmonds, 'Fossil Collection of the Misses Philpot', p. 44.
37. *Ibid.*
38. Lang, 'Mary Anning and the Pioneer Geologists', p. 56.
39. Lang, 'Mary Anning and Anna Maria Pinney', p. 149.
40. The whole of Sir Philip Egerton's collection was purchased by the Trustees of the British Museum in 1882; the collection of William Willoughby Cole, Lord Cole (Earl of Enniskillen), was purchased in 1882 and 1883, including the *Plesiosaurus macrocephalus* found by Mary Anning.
41. Gordon, *Life and Correspondence of William Buckland*, p. 139.
42. Letter, 7 April 1839, *Magazine of Natural History*, NS3 (1839), 605, quoted in Lang, 'Mary Anning and the Pioneer Geologists', p. 154.
43. Lang, 'Mary Anning and the Pioneer Geologists', p. 156.
44. Roberts, *History of Lyme Regis*, p. 324.
45. *Ibid.*, p. 325.
46. Duncan was later elected keeper of the Ashmolean Museum, Oxford, in 1823, and did much to reorganise the exhibits and introduce a reference library.
47. Nicolaas Rupke, *The Great Chain of History: William Buckland and the English School of Geology (1814–1849)*, Oxford, Oxford University Press, 1983, p. 142.
48. Lang, 'Mary Anning and the Pioneer Geologists', p. 155.
49. Edmonds, 'Fossil Collection of the Misses Philpot', p. 44.
50. *Transactions of the Geological Society of London*, 2nd series, 3 (1829), pp. 217–22, quoted in Roberts, *History of Lyme Regis*, pp. 325–6.
51. Fragments of specimen; not on public display.
52. NHM Ref. No. R 1034.
53. Taylor and Torrens, 'Saleswoman', p. 135.
54. Price, 'Mary Anning Specimens', p. 320.

55. Taylor and Torrens, 'Saleswoman', p. 138.
56. *Ibid.*, p. 140.
57. Lang, 'Mary Anning and the Pioneer Geologists', p. 155.
58. Richard Owen, 'A Description of the Plesiosaurus macrocephalus, Conybeare, in the Collection of Viscount Cole', *Transactions of the Geological Society of London*, 2nd series, 5 (1840), pp. 515–35 and plates 43–5. Torrens, 'Presidential Address', p. 267.
59. NHM Ref. No. R 1158.
60. Murray, 'Late Miss Mary Anning', p. 591.

Chapter Four
1. Stephen Jay Gould, *Finders, Keepers: Eight Collectors*, London, Hutchinson, 1992, p. 109.
2. J. Clark, *Memoirs*, Street, privately printed, 1920.
3. Lang, 'Mary Anning and the Pioneer Geologists', p. 156.
4. W.D. Lang, 'Mary Anning, of Lyme, Collector and Vendor of Fossils, 1799–1847', *Natural History Magazine*, 5 (1936), p. 77.
5. Hawkins, *Memoirs of Ichthyosauri and Plesiosauri*, p. ix.
6. Lang, 'Mary Anning of Lyme, Collector and Vendor', p. 77.
7. Hawkins, *Memoirs of Ichthyosauri and Plesiosauri*, pp. 12–13.
8. *Ibid.*, pp. 25–7.
9. *Ibid.*, pp. 9–10.
10. Lang, 'Three Letters', pp. 169–73.
11. Lang, 'Mary Anning and the Pioneer Geologists', p. 159.
12. Lang, 'Mary Anning and Anna Maria Pinney', p. 147.
13. Lang, 'Mary Anning and the Pioneer Geologists', p. 159.
14. All Bell references, list and exchange of letters from Johnson, *A Memoir of Miss Frances Augusta Bell*.
15. Lang, 'Mary Anning and Anna Maria Pinney', pp. 146–52.
16. John Fowles, *Three Town Walks*, Lyme Regis Philpot Museum, 1983, inside front cover; and Roberts, *History of Lyme Regis*, p. 56.

17. Miss Marigold Watney, 'Mary Anning – First Woman Geologist', *Morning Post*, 4 February 1931, quoted in Lang, 'Mary Anning and the Pioneer Geologists', p. 144.
18. Roberts, *History of Lyme Regis*, p. 174.
19. All Bell references and exchange of letters from Johnson, *A Memoir of Miss Frances Augusta Bell*.
20. John L. Morton, *King of Siluria: How Roderick Murchison Changed the Face of Geology*, Horsham, Broken Spectre Publishing, 2004, p. 83.
21. Lang, 'Three Letters', p. 169.
22. *Ibid.*, p. 170.
23. Torrens, 'Presidential Address', p. 267.
24. Lang, 'Mary Anning and Anna Maria Pinney', p. 147.
25. Jo Draper, *Mary Anning's Town, Lyme Regis*, Lyme Regis Philpot Museum, 2004, p. 29.
26. All references to London Diary: Richard Owen Correspondence: General Library, Natural History Museum, vol. 1, pp. 149–50. The diary was purchased by Lord Enniskillen, from Mary's nephew, Albert Anning. There is also a letter to Owen by Lord Enniskillen, and several other transcripts in her writing.
27. Archibald Geikie, *Life of Sir Roderick I. Murchison*, London, John Murray, 1875, vol. 1, p. 128.
28. Lang, 'Three Letters', pp. 169–70.
29. Charles Dickens, 'Mary Anning, the Fossil Finder', *All the Year Round*, 13 (1865), p. 62.
30. The Reform Act of 1867 abolished this constituency, which was absorbed into a larger district.
31. Cyril Wanklyn, *Lyme Leaflets*, London, Spottiswoode, Ballantyne & Co. Ltd, 1944, p. 84.
32. 'Lister's Thesaurus' quoted in Lang, 'Mary Anning and Anna Maria Pinney', pp. 146–52.
33. [J.S. Buckingham and James Silk], *A Summer Trip to Weymouth and Dorchester . . .*, 1842. Copy in Dorset County Library.

34. The railway reached Lyme in 1903 and retreated to Axminster in 1965.
35. David Price, 'Mary Anning Specimens in the Sedgwick Museum, Cambridge', *Geological Curator*, 4/6 (1986), p. 322.

Chapter Five

1. Molly Anning referring to her daughter Mary; *Chambers's Journal* (1857), p. 314.
2. Lang, 'Three Letters', p. 171.
3. Lang, 'Mary Anning and the fire', p. 175.
4. 'X', *Lyme 50 Years Ago*.
5. Lang, 'Mary Anning and Anna Maria Pinney', pp. 146–52.
6. *Ibid.*, p. 148.
7. Roberts, *History of Lyme Regis*, p. 290.
8. E. Welch, 'Lady Silvester's Tour through Devonshire in 1824', *Devon and Cornwall Notes and Queries*, 30 (1967), p. 313; 32 (1967), pp. 265–6.
9. Price, 'Mary Anning Specimens', p. 322.
10. Lang, 'Three Letters', p. 169.
11. Murray, 'Late Miss Mary Anning', p. 591.
12. *Chambers's Journal*, 1857, p. 383.
13. Lang, 'Mary Anning and Anna Maria Pinney', pp. 146–52.
14. Dickens, 'Mary Anning, the Fossil Finder', p. 62.
15. Roberts, *History of Lyme Regis*, p. 290.
16. Lang, 'Mary Anning and Anna Maria Pinney', pp. 146–52.
17. Richard Owen correspondence, vol. 1, pp. 150–2.
18. W.L. Rowles, *Days Departed; or Banwell Hill*, London, J. Murray and Bath, 1829.
19. Lang, 'Mary Anning and Anna Maria Pinney', p. 148.
20. 'No. 4' notebook, Lang Papers, Dorset County Museum, Dorchester (NHMS XXXVII/2).
21. Lang, 'Mary Anning and Anna Maria Pinney', p. 150.
22. Waring, *Peeps*, p. 4.
23. *The Complete Works of Henry Kirke White*, ed. Robert

Southey, Boston, Whitaker, 1831, in 'No. 4' notebook, Lang Papers (NHMS XXXVII/2).

24. Gordon, *Life and Correspondence of William Buckland*, p. 123.
25. *Ibid.*, p. 147.
26. *Ibid.*, p. 113.
27. Lang, 'Mary Anning and Anna Maria Pinney', p. 147.
28. Lang, 'Three Letters', p. 171.
29. Lang, 'Mary Anning and Anna Maria Pinney', p. 148.
30. *Ibid.*
31. *Ibid.*, p. 149.
32. Edward Pidgeon, *Fossil Remains of the Animal Kingdom*, London, Whitaker, 1830, p. 377, quoted in Lang, 'Mary Anning and Pioneer Geologists', p. 150.
33. *Bristol Mirror*, 11 January 1823, p. 4, quoted in Torrens, 'Presidential Address', pp. 257–84.
34. Lang, 'Three Letters', p. 170.
35. 'X', *More Memories of Old Lyme*.
36. Lang, 'Three Letters', p. 170.
37. Edmonds, 'Fossil Collection of the Misses Philpot', p. 45.
38. Roberts, *History of Lyme Regis*, p. 287; Lang, 'Mary Anning's Escape from Lightning', pp. 91–3.
39. W.D. Lang, 'Mary Anning's escape from lightning', *Proceedings of the Dorset Natural History and Archaeological Society*, 80 (1959), p. 91.
40. Lang, 'Mary Anning and Anna Maria Pinney', p. 146.
41. Roberts, *History of Lyme Regis*, pp. 287–8.
42. Murray, 'Late Miss Mary Anning', p. 591.
43. Roberts, *History of Lyme Regis*, p. 172.
44. Jane Austen, *Letters*, ed. Le Faye, p. 144.
45. Maggie Lane, *Jane Austen and Lyme Regis*, Chawton, Jane Austen Society, 2003, pp. 38–40.
46. Lang, 'Mary Anning and the fire', p. 175. The bell from the clock tower is in the museum.

47. *Extraordinary Land Slip and Great Convulsion of the Coast of Culverhole Point, near Lyme Regis, Dorset*, Lyme Regis, Dunster's General Printing Office, 1839.

48. John Vaughan, 'Jane Austen at Lyme', *Monthly Packet*, 4th series, 6 (1893), pp. 271–9.

49. W.D. Lang, 'More About Mary Anning, including a Newly-Found Letter', *Proceedings of the Dorset Natural History and Archaeological Society*, 71 (1949), p. 185.

50. M. Aurousseau (ed.), *The Letters of F.W. Ludwig Leichhardt*, 3 vols, Cambridge, Cambridge University Press, 1968, vol. 1, p. 232.

51. Carus, *King of Saxony's Journey*, p. 197.

52. Torrens, 'Presidential Address', p. 279.

53. *Ibid.*, p. 55.

54. Edmonds, 'Fossil Collection of the Misses Philpot', p. 46.

55. James Sowerby, *Mineral Conchology of Great Britain, 1812–32*, vol. 6, p. 96.

56. Torrens, 'Presidential Address', pp. 281–2.

Chapter Six

1. Gould, *Finders Keepers*, p. 100.

2. Mantell MSS, Alexander Turnbull Library, Wellington, New Zealand, quoted in Torrens, 'Presidential Address', p. 10.

3. Roberts, *History of Lyme Regis*, opposite p. 290; copy annotated by the author, in Lyme Regis Philpot Museum.

4. Murray, 'Late Miss Mary Anning', p. 391.

5. Minutes of the Dorset County Museum, Dorchester, 21 July 1846.

6. Letter from William Buckland to Sir Walter Trevelyan, 10 October 1846, British Library, Add. MSS 31026/247.

7. Roberts, *History of Lyme Regis*, p. 178.

8. (Fanny Burney), *Diary and Letters of Madame D'Arblay*, London, Swan Sonnenschein & Co., 1893.

9. Lang, 'Mary Anning and the Pioneer Geologists', p. 162.

10. Warner and Gupta, *Bennetts of Lyme Regis*, pp. 22, 54, 103.

11. Lang, 'Mary Anning and the Pioneer Geologists', p. 149.
12. Torrens, 'Presidential Address', p. 278.
13. A.G. Mantell, 'A Few Notes on the Prices of Fossils', *The London Geological Journal*, No. i, September 1846, pp. 13–17, quoted in Lang, 'Mary Anning and the Pioneer Geologists', p. 151.
14. Gordon, *Life and Correspondence of William Buckland*, p. 24.
15. *Dorset County Chronicle*, 27 August 1857.
16. Lang, 'Mary Anning and the Pioneer Geologists', p. 154.
17. F. Bickley, *Where Dorset Meets Devon*, London, Constable, 1911, p. 55.
18. Roberts, *History of Lyme Regis*, p. 290.
19. J.W. Preston, *Lyme Lyrics*, 1884.
20. Letter, 23 November 1942, Lang; 'More about Mary Anning', p. 188.
21. The 'greatest fossilist' annotation is on an undated letter from Mary Anning to one of the Misses Philpot of Lyme, in the collection of the American Philosophical Society, Philadelphia, cited in Hugh Torrens, 'Mary Anning (1799–1847) of Lyme: "the greatest fossilist the world ever knew"', (Presidential Address), *British Journal for the History of Science*, 28 (1995), pp. 257–84.
22. John Kenyon, *Poems: For the Most Part Occasional*, London, 1838, pp. 109–11.

Appendix
1. Roberts, *History of Lyme Regis*, pp. 320–1.
2. Hugh Torrens, 'E.T. Higgins (*c.* 1816–1891), Geological Curator and Natural History Dealer', in D.F. Brannagan and G.H. McNally (eds), *Useful and Curious Geological Enquiries*, Sydney, International Commission on the History of Geological Sciences, 1994, pp. 200–8, quoted in Goodhue, *Fossil Hunter*, p. 116.

Bibliography

MANUSCRIPTS

CANADA
Montreal
McGill University, The Blacker–Wood Collection of Rare Books
and Special Collections Division
MA letter to Edward Charlesworth
Two pages of MA's pensées, and letter from C. Davies Sherborn
regarding their provenance

NEW ZEALAND
Wellington
Alexander Turnbull Library, National Library of New Zealand,
Te Puna Matauranga o otearoa
Mantell MSS

UNITED KINGDOM
Bristol
Bristol University Library
G. Roberts, 'A Brief Memoir of Miss Mary Anning, the
Celebrated Fossilist', Eyles MSS, 1847, unpublished

City of Bristol Museum and Art Gallery
MA letter to J.W. Miller of 20 January 1830

Cambridge
Cambridge University Library
Add. MS 7652, 7918, No. 5064: eleven Mary Anning letters,
    including Add. MS 7652/II, LL 21C, 23 and KK 11
Sedgwick re *Squaloraja*
Sedgwick's 'rough' account book
Woodwardian accounts

Sedgwick Museum of Earth Sciences
Sedgwick's field journal (Journal No. 5) for September
    1820

Cardiff
National Museum of Wales
Geology Department: De La Beche Archives, including
    Conybeare's letter to De La Beche (1821) (NMW
    84.20G.D.299) and Murchison's letter to De La Beche (NMW
    84.29G.D.1020)
Zoology Department: G.B. Sowerby Archives 329

Devizes
Wiltshire Archaeological Society Library
Cunningham MS: James Johnson letter to Cunningham, 23 July
    1810

Dorchester
Dorset County Museum
Owen Archives
Mary Anning, 'No. 4' notebook', 1840s, and other papers
    (NHMS XXXVII/2)

Dorset County Record Office
PE/LROV6: Overseers of the Parish Poor

London
British Library
Add. MS 36520: Cumberland Papers (letter 20 August 1820, MA 'attends' Birch)
Add. MS 31026

British Museum
In volume C 1467 (letter wrongly dated 1820): De La Beche letter to Keeper

Natural History Museum
Earth Sciences Library
MSS ANN 1: Manuscript and Drawing Collection of Mary Anning
Archives – NS: Additions, Geology, 12001–16000,73
'Notebook of Thomas Allen: Travels in England, 1813–1814, MS, 1824
'A Catalogue of a small but very fine collection of organized fossils, from the Blue Lias Formation, At Lyme and Charmouth, in Dorsetshire, consisting principally of Bones, Illustrating the Osteology of the Icthio-saurus, or Proteo-saurus, and of specimens of the Zoophyte, called Pentacrinite, the genuine property of Colonel Birch, Collected at considerable expense, which will be sold by auction by Mr Bullock, at his Egyptian Hall in Piccadilly, on Monday, the 15th day of May, 1820'
General Library
I.149C: Richard Owen Correspondence
OC 62.1: Owen MSS

Public Record Office
RG 4/462: Independents Chapel (Coombe Street Chapel) register

Lyme Regis
Dinosaur Museum
MS Independents Chapel 'Church Book', 1852–6, Revd S. Knell

USA
Philadelphia
American Philosophical Society
Letter from MA to Miss Philpot, undated

BOOKS

Austen, Jane, *Letters*, ed. Deidre Le Faye, Oxford, Oxford
  University Press, 1995

Austen, Jane, *Persuasion*, London, Penguin, 1998

Brown, H. Roland, *The Beauties of Lyme Regis, Charmouth, the
  Landslip and their Vicinities*, London, Longman, Brown,
  Green and Longmans, 1857; reprinted Lyme Regis, Dunster,
  Lymelight Press, 1997

[Buckingham, J.S., and Silk, James], *A Summer Trip to
  Weymouth and Dorchester, Including an Excursion to
  Portland, and a Visit to Maiden Castle, and the
  Amphitheatre, From the Notebook of an Old Traveller*
  [J.S. Buckingham] 1842

Buckland, William, 'Introduction', in *Geology and Mineralogy
  Considered with Reference to Natural Theology* (Bridgewater
  Treatise), London, William Pickering, 1836

Cadbury, Deborah, *The Dinosaur Hunters*, London, Fourth
  Estate, 2000

Carus, C.G., *The King of Saxony's Journey through England and
  Scotland in the Year 1844*, London, Chapman & Hall, 1846

Chambers's Journal, London and Edinburgh, William
  Chambers, 1857

Clark, J., *Memoirs*, Street, privately printed, 1920

Curle, R., *Mary Anning 1799–1847*, Dorchester, Dorset Natural
  History and Archaeological Society (Dorset Worthies, No. 4),
  1963

Dictionary of National Biography, Oxford, Oxford University
  Press, 2005

Draper, Jo, *Mary Anning's Town, Lyme Regis*, Lyme Regis
  Philpot Museum, 2004

Fowles, John, *Three Town Walks*, Lyme Regis, Friends of the
Lyme Regis Philpot Museum, 1983

Fowles, John, *A Short History of Lyme Regis*, Wimborne,
Dovecote Press, 1991

Geikie, Archibald, *Life of Sir Roderick I. Murchison*, London,
John Murray, 1875

Goodhue, Thomas, *Fossil Hunter: The Life and Times of Mary
Anning, 1799–1847*, Bethesda, Md, Academica Press,
2004

Gordon, Elizabeth Oke Buckland, *The Life and Correspondence
of William Buckland, DD, FRS, Sometime Dean of
Westminster, Twice President of the Geological Society, and
First President of the British Association*, London, John
Murray, 1894

Gould, Stephen Jay, and Purcell, Rosamund Wolfe, *Finders,
Keepers: Eight Collectors*, New York, Norton, 1992; London,
Hutchinson, 1992

Hack, Maria, *Geological Sketches and Glimpses of Ancient
Earth*, London, Harvey & Darton, 1832

Hallett, Selina, *Lyme Voices*, 1, with Foreword and Notes by
John Fowles, Lyme Regis Philpot Museum, 1991

Hawkins, Thomas, *Memoirs of Ichthyosauri and Plesiosauri:
Extinct Monsters of the Ancient Earth, with Twenty-One
Plates Copied from Specimens in the Author's Collection*,
London, Relfe & Fletcher, 1834

Hill, Constance, *Jane Austen: Her Homes and her Friends*,
London, Bodley Head, 1923

Hutton, James, *Theory of the Earth with Proofs and
Illustrations*, Edinburgh, Messrs Cadwell, Junior, Davies and
W. Creech, 1795, vol. 1

Jefferis, B.G., and Nichols, J.L., *The Household Guide or
Domestic Cyclopedia*, Toronto, Nichols, 1894

Johnson, Grant, *A Memoir of Miss Frances Augusta Bell who
Died in Kentish Town, Monday, 23rd of May, 1825, Aged
Fifteen Years and Six Months; with Speciments of her*

Compositions, in Prose and Verse, London, Hatchard & Son, 1827

Lane, Maggie, *Jane Austen and Lyme Regis*, Chawton, Jane Austen Society, 2003

Leichhardt, Ludwig, *The Letters of F.W. Ludwig Leichhardt*, collected and newly translated, ed. M. Aurousseau, 3 vols, London, Cambridge University Press for the Hakluyt Society, 1968

McCartney, Paul J., *Henry De La Beche: A New Kind of Geologist*, Cardiff, Friends of the National Museum of Wales, 1977

McGowan, Christopher, *The Dragon-Seekers: The Discovery of Dinosaurs during the Prelude to Darwin*, London, Little, Brown/Abacus, London, 2002

Malone, Edward, *An Account of the Incidents from which the Title and Part of the Story of Shakespeare's Tempest were Derived, and its True Date Ascertained*, London, published privately, 1808

Mantell, Gideon, *The Journal of Gideon Mantell, Surgeon and Geologist: Covering the Years 1818–1852*, ed. E. Cecil Curwen, Oxford, Oxford University Press, 1940

Mantell, Gideon, *Thoughts on a Pebble, or A First Lesson in Geology*, Reeve, Benham & Reeve, London, 1849

Morton, John L, *King of Siluria: How Roderick Murchison Changed the Face of Geology*, Horsham, Broken Spectre, 2004

Munby, Lionel, *How Much Is That Worth?*, 2nd edn, Chichester, Phillimore for British Association for Local History, 1996

North, The Hon. Roger, *The Life of the Right Honourable Francis North, Baron of Guilford, Lord Keeper of the Great Seal, under King Charles II and King James II. Wherein are Inserted the characters of Sir Matthew Hale, Sir George Jeffries, Sir Leoline Jenkins, S. Godolphin and others, the Most Eminent Lawyers and Statesmen of that Time*, 3 vols, London, Henry Cobham, 1742

Parsons, J.F., *Princess Victoria in Dorset*, Bournemouth, Higher Studies, 1996

Parker, Louisa, Ford, Judy and Draper, Jo, *Ethnic Minorities in Lyme Regis and West Dorset Past and Present*, Lyme Regis, Philpot Museum, 2004

Preston, J.W., *Lyme Lyrics*, Lyme Regis, 1884

Roberts, George, *Roberts's History of Lyme Regis and Charmouth*, London, Samuel Bagster & William Pickering, 1834

Roberts, George, *An Etmological and Explanatory Dictionary of the Terms and Language of Geology*, London, Longman, Orne, Brown, Green and Longmans, 1839

Roberts, George, *A Social History of the People of the Southern Counties of England in Past Centuries*, London, Longman, Brown, Green, Longmans & Roberts, 1856

Rowles, W.L., *Days Departed; or Banwell Hill*, London, J. Murray and Bath, 1829

Rupke, Nicolaas H., *The Great Chain of History*, Oxford, Oxford University Press, 1983

Southgate, Henry, *Things a Lady Would Like to Know*, London and Edinburgh, William P. Nimmo, 1875

Steinbach, Susie, *Women in England, 1760–1914: A Social History*, London, Weidenfeld & Nicolson, 2004

Tickell, Crispin, *Mary Anning of Lyme Regis*, Lyme Regis, Philpot Museum, 1996

Vallone, Lynne, *Becoming Victoria*, New Haven and London, Yale University Press, 2001

Wanklyn, Cyril, *Lyme Leaflets*, London, Spottiswoode, Ballantyne & Co. Ltd, 1944

Wanklyn, Cyril, *Lyme Regis: A Retrospect*, London, Arthur L. Humphreys, 1922

Waring, Nellie (Sister Emma) (S.E.), *Peeps into an Old Playground: Memories of the Past*, Lyme Regis, Dunster, 1895

Warner, Jill, and Gupta, Pam Bennett (researcher), *The Bennetts of Lyme Regis, 1762–1911*, Wimborne, Dovecote Press, 1997

'X', *Lyme 50 Years Ago*, Lyme Regis, 1892
'X', *More Memoires of Old Lyme*, Lyme Regis, 1892

MAGAZINES AND JOURNALS

Buckland, William, 'On the Discovery of a New Species of Pterodactyle in the Lias at Lyme Regis', *Transactions of the Geological Society of London*, 2nd series, 3 (1829), pp. 217–22

Buckland, William, 'On the Discovery of Coprolites, or Fossil Faeces, in the Lias at Lyme Regis, and in Other Formations' (read 6 February 1829), *Transactions of the Geological Society of London*, 2nd series, 3 (1835), pp. 221–36

De La Beche, H., and Conybeare, W.D., 'Notice of the Discovery of a New Fossil Animal', *Transactions of the Geological Society of London*, 5 (1821), pp. 559–94

Dickens, Charles, 'Mary Anning, the Fossil Finder', *All the Year Round*, 13 (1865), pp. 60–3

Dorset County Chronicle, 27 August 1857

Edmonds, J.M., 'The Fossil Collection of the Misses Philpot of Lyme Regis', *Proceedings of the Dorset Natural History and Archaeological Society*, 98 (1976), pp. 43–8

Fowles, John, 'Curator's Report for the Lyme Regis Museum', Lyme Regis Philpot Museum, 1993–4

Goodhue, Thomas W., 'The Faith of a Fossilist: Mary Anning', *Anglican and Episcopal History*, 70/1 (2001), pp. 80–100

Lang, W.D., 'Mary Anning of Lyme, Collector and Vendor of Fossils, 1799–1847', *Natural History Magazine*, 5 (1936), pp. 64–81

Lang, W.D., 'Demonstration at the British Museum (Natural History), South Kensington, "Mary Anning, Fossilist"', *Proceedings of the Geologists' Association of London*, 47/1 (1936), pp. 65–7

Lang, W.D., 'Mary Anning and the Pioneer Geologists of Lyme', *Proceedings of the Dorset Natural History and Archaeological Society*, 60 (1939), pp. 142–64

Lang, W.D., 'Three Letters of Mary Anning, "Fossilist" of Lyme',
   *Proceedings of the Dorset Natural History and
   Archaeological Society*, 66 (1945), pp. 169–73
Lang, W.D., 'More about Mary Anning, Including a Newly-
   Found Letter', *Proceedings of the Dorset Natural History and
   Archaeological Society*, 71 (1950), pp. 184–8
Lang, W.D., 'Mary Anning and the Fire at Lyme in 1844',
   *Proceedings of the Dorset Natural History and Archaeo-
   logical Society*, 74 (1953), pp. 175–7
Lang, W.D., 'Mary Anning and Anna Maria Pinney',
   *Proceedings of the Dorset Natural History and Archaeo-
   logical Society*, 76 (1955), pp. 146–52
Lang, W.D., 'Mary Anning's Escape from Lightning',
   *Proceedings of the Dorset Natural History and Archaeo-
   logical Society*, 80 (1959), pp. 91–3
Lang, W.D., 'Portraits of Mary Anning and Other Items',
   *Proceedings of the Dorset Natural History and Archaeo-
   logical Society*, 81 (1960), pp. 89–91
Lang, W.D., 'Mary Anning and a Very Small Boy', *Proceedings
   of the Dorset Natural History and Archaeological Society*, 84
   (1963), pp. 181–2 (privately printed in *Family Quartets* by
   Cosmo W. Rawlins)
Murray, J., 'The Late Miss Mary Anning', *Mining Journal*, 17, 25
   December 1847
Owen, Richard, 'A Description of the *Plesiosaurus
   macrocephalus*, Conybeare, in the Collection of Viscount
   Cole' (read 4 April 1838), *Transactions of the Geological
   Society of London*, 2nd series, 5 (1840), pp. 515–35
Playfair, J., 'Biographical Account of the Late James Hutton,
   FRS Edinburgh', *Transactions of the Royal Society of
   Edinburgh*, 5/3 (1805)
Price, David, 'Mary Anning Specimens in the Sedgwick
   Museum, Cambridge', *Geological Curator*, 4/6 (1986),
   pp. 319–24

Taylor, Michael A, 'Collections, Collectors and Museums of
   Note: The Lyme Regis (Philpot) Museum, the History,
   Problems and Prospects of a Small Museum and its
   Geological Collection', *Geological Curator*, 4 (1986),
   pp. 309–17

Taylor, M., and Torrens, Hugh S., 'Saleswoman to a New
   Science: Mary Anning and the Fossil Fish Squaloraja from
   the Lias of Lyme Regis', *Proceedings of the Dorset Natural
   History and Archaeological Society*, 108 (1986), pp. 161–8

Torrens, Hugh S., 'Collections and Collectors of Note: 28
   Colonel Birch (*c.* 1728–1829)', *Geological Curator*, 2 (1979),
   pp. 405–12

Torrens, Hugh S., 'Presidential Address: Mary Anning (1799–
   1847) of Lyme: "The Greatest Fossilist the World Ever
   Knew"', *British Journal for the History of Science*, 28 (1995),
   pp. 257–84

Torrens, Hugh, 'Mary Anning (1799–1847) Notes for a Historical
   Tour in her Footsteps around Lyme Regis', Mary Anning
   Symposium, Lyme Regis Philpot Museum, June 1999

Welch, E., 'Lady Silvester's Tour through Devonshire in 1824',
   Devon and Cornwall Notes and Queries, 32 (1973),
   pp. 255–66

Index

Kent, Duchess of 144
Kenyon, John 195
Kirkdale Cavern 104, 105
Knell, Revd S. 223
Konig, Charles 19, 138

Lamarck, Jean Baptiste de 149
Lambert Castle, Dorset 3
Landslip of 1839 172–5
Lang, William Dickson 16, 115
Leichhardt, Friedrich Ludwig 175
*Leptopterygius platyodon, see
    Ichthyosaurus*
Lever, Sir Ashton 110
liassic fish 102
Lister, Arthur 70
Lister, Guilelma 69–70, 193, 210
Lister, Joseph 70
Lister, Miss M. 174
Llhyd 2
Lock, William 13
London 21–2, 84, 85, 86, 92,
    107, 110, 132–40, 150, 188–9
London Bridge 134–5
'London Museum' *see* Bullock
Lonsdale, William 36
Lord, Alan 180
Lyceum of Natural History, New
    York 96
Lyell, Charles 48, 53, 95, 111,
    126, 138
Lyell, Mary (*née* Horner) 95
Lyme Regis, Dorset 2, 134;
    Assembly Rooms 11, 33, 36, 82,
    142, 166, 187; baths 7, 33, 42;
    Black Ven cliff 3, 12, 17, 19, 69,
    80, 82, 92, 197; bread riot 7; civil
    war siege 4; Cobb 3, 8, 12, 31,
32, 35, 91, 92, 93, 129, 145;
    Cockmoile Square 6, 10–11, 93,
    96, 112, 145, 168, 171, 186;
    Customs 8; fires 5–6, 171–2;
    Fossil Depot (MA's) 21, 187;
    Fossil Depot (Curtis's) 186–7;
    fossiling 13, 22, 23; Great
    Landslip 172–5; Independents'
    (Congregationalist) Chapel (now
    Dinosaur Museum) 8, 48, 65–6,
    25, 223; infant mortality 6;
    Monmouth Beach 5, 69; railway
    147, 216; Rotten Borough 67,
    142–3; St Michael the Archangel
    12, 16, 17, 30, 78, 116, 186, 192,
    193; sea bathing 36, 142; seaside
    resort 5, 28–31, 33, 35–7;
    shipbuilding 93; storms 128–30;
    Great Storm of 1824 91, 128;
    Undercliff 4, 172–3; walk in
    Mary's day 10–12
Lyme Regis Philpot Museum
    187, 194, 201, 212

McAdam 42
McGill University, The Blacker-
    Wood Collection of Rare
    Books and Special
    Collections Division,
    Montreal 220
Malone, Edmond 34
Manchester 142
Mantell, Gideon Algernon 1, 24,
    39, 59–61, 64, 65, 80, 98, 99,
    104, 121, 123, 124, 125, 126,
    141, 153, 178, 188, 192, 220
Mantell, Mrs Mary 60
marcasite 10